Bowyer, P. J. (Peter J.)

Trouble shooting and maintenance of boat engines

DATE			

Trouble shooting and maintenance of boat engines

Trouble shooting and maintenance of boat engines

Peter Bowyer

NAUTICAL BOOKS
MACMILLAN LONDON

Acknowledgements

The principal artwork was drawn by Dennis Fairfield MSIAD.
Author's sketches have been traced by Diane Wagstaffe.

The author gratefully acknowledges the help given by the following companies in providing information and illustrations.

American Bosch
A.R.S. Marine Limited
BMW Marine GMBH
E. J. Bowman (Birmingham) Limited
Carl Hurth GMBH
The Chloride Group Limited
AC Delco Division of General Motors Corporation
Detroit Diesel Allison, Division of General Motors Corporation
M. G. Duff & Partners Limited
Golden Arrow (Extinguishers) Limited
Gunson's Colorplugs Limited
Jabsco Products – ITT Corporation
Lucas Electrical Limited
Lucas–CAV Limited
Lucas Marine Limited
Mercury Marine
Motorola Limited
Perkins Engines Limited
Robert Bosch GMBH
SU Fuel Systems
Teignbridge Engineering Limited
Teleflex Morse
Warner Gear Division of Borg-Warner Corporation
Watermota Limited

First published in Great Britain 1983 by
NAUTICAL BOOKS
an imprint of Macmillan London Ltd
4 Little Essex Street
London WC2R 3LF

Associated companies throughout the world

ISBN 0 333 345568

Typeset by Wyvern Typesetting Ltd, Bristol

Printed at The Pitman Press, Bath

Contents

Introduction

Most people running a boat – whether for pleasure or as a business – also use a car and, in these days of rising garage costs, more and more drivers are doing their own servicing and in many cases engine 'tune-ups' and repairs.

Whether the boat engine runs on diesel or petrol fuel, many of its external components and the way it is installed are different from the car engine's. This, coupled with a fear of causing the engine to fail at sea, has put off many would-be maintenance men, who as a consequence never get beyond changing the lubricating oil in the sump.

Understanding how the engine works, and what it needs to keep going just when you most depend on it, will enable you to maintain it yourself and also give you the know-how to deal with any problems that come up, so that you can 'trouble-shoot' with confidence.

Keeping the engine in tune can also save money – not only because you are doing the work yourself but because you will be squeezing all the available power out of each drop of expensive fuel.

Many of the problems occurring with boat engines are caused by the marine environment and the intermittent nature of the operation. You climb into the car most days and run the engine up to its full working temperature, even if you are only doing a few miles to the office and back. Except for a few weeks or long weekends in the summer, the pleasure-boat engine will lie unused, if not neglected, for long periods if it is an auxiliary, and the short time that you are motoring at half-throttle until you can hoist the sails and switch off will not give it its necessary 'constitutional'.

If yours is a sports-boat, cruiser or working vessel the engine has a better chance of being 'warmed through' each time it is used and it will respond to this by keeping in good condition for much longer, with less tendency to produce sludge in the oil sump and other areas. Many engines run more sweetly when they are properly warmed up – you can hear the engine note change, and this applies to diesel as well as petrol models.

In fact there are many similarities between diesel and petrol engines installed in boats, which make some maintenance procedures common to both types. Although the principal components of the diesel are massive because of the higher firing pressures produced in the cylinders, many parts are indistinguishable from the petrol counterpart, and you don't need a Chief Engineer's certificate to maintain a diesel engine. We hope this book will persuade you to have a go at keeping yours in good condition.

'Preventive maintenance' is the approach favoured by all the engine manufacturers. A regular routine which you can establish yourself, based on the recommendations given here, will reveal many potential problems before they develop into real trouble, as well as saving money on expensive repairs and extending the working life of 'old faithful'.

With many boats there are occasions during a cruise when shortcomings of the engine installation become apparent. A need for better soundproofing is often expressed, and improved vibration insulation may become a priority after hours of motoring the engine. These jobs and many others can best be tackled when the boat is laid up for the winter, and, whether the boatyard is commissioned or a DIY approach is decided on, a certain amount of planning is called for.

Safety at sea includes guarding against accidents on-board, and here of course we should not wait until the end of the season to provide adequate fire-extinguishers, or to fit guards over dangerous parts of the machinery. Making these provisions, as well as having the know-how to deal with any mechanical problems, will help to boost confidence.

The Achilles' heel of a diesel engine is the fuel system and that of a petrol engine is the electrical system. The importance of maintaining both correctly is dealt with in some detail, and the benefit of good installation practice is explained.

The battery is vital to electric-starting diesel and petrol engines, and the combination of long periods of disuse, corrosion from damp, a salty atmosphere and heavy demand for short periods is not conducive to long, trouble-free life. The battery is another example of getting what you pay for; sea-going conditions call for a marine battery, and the difference between this and automotive batteries is described, as well as precautions in its use and good maintenance practice.

Corrosion can be one of the worst problems on a boat; it is often caused by the use of incompatible metals as well as unsuitable ones, such as brass, which de-zincifies in the presence of sea water although it is perfectly adequate in fresh. These problems often arise when you are carrying out repairs or adding fittings to a boat; guidelines for avoiding corrosion are given.

Reading this preamble may give the impression that looking after a boat engine is a difficult and time-consuming task. This is not so, especially if opportunities to improve the installation and accessibility for servicing are taken. Much of the information here will enable routine tasks and improvements to be carried out without the specific workshop manual for the engine. However, it is always advisable to get hold of an owner's guide if you do not already possess one, so that maintenance tasks can be carried out at the correct intervals and any peculiarities of your particular power unit can be revealed.

Boat-owners' service courses are to be recommended. Engine-manufacturers offering short courses on a particular engine model can impart servicing procedures and know-how much more easily than through the pages of their manuals. The latter serve as an *aide-mémoire*, however, and should always be kept on board for reference.

This book is not just about the engine, which, after all, is simply the means of turning a shaft to which a rather expensive piece of twisted metal is attached – and if this gadget is damaged, misshapen or even wrongly specified, all our efforts to keep the power unit in tip-top condition are wasted. The reliability of 'peripherals such as the reverse gearbox is obviously of vital importance. So also are engine support systems such as the fuel supply, cooling water, air intake, exhaust piping etc. Some items play a dual role – in most craft the thrust developed by the propeller is transmitted to the engine via the propeller-shaft and gearbox, and it is the holding-down bolts of the engine that finally pass this force to the hull and actually propel the boat.

1 How it works

It is certainly not necessary to be an expert engineer to carry out routine maintenance on your boat engine – or in fact to 'trouble-shoot' when problems arise. However, a good grounding in the basic principles of 'how it works' will provide a useful foundation of confidence when you tackle any job on the engine.

In this chapter we take a look inside petrol and diesel engines, noting their similarities and differences, and then examining the ways in which they are adapted for use in boats. I make no excuse for starting from basic principles because, although many readers will have lived for years with engines in cars and possibly commercial vehicles as well as boats, perhaps some point was never clear and can be clarified in the following pages.

Petrol and diesel engines are built with from one to eight or more cylinders, but the four-cylinder, *four-stroke* engine is probably the most common, so we shall use a petrol engine of this type as our example. (Plate 1.1)

This engine has its cylinders arranged *in line*; other engines – mainly six- and eight-cylinder – have them in vee formation. Less commonly there are engines with cylinders horizontally opposed. (Figure 1.1)

The central mechanical feature of the engine is the crankshaft, to which the pistons are linked by connecting rods. Downward movement of the pistons causes the crankshaft to rotate so that power is transmitted to the flywheel, then to a gearbox, and finally to the propeller-shaft and propeller. (Figure 1.2)

The four 'strokes', i.e. phases of the cycle, are performed as follows:

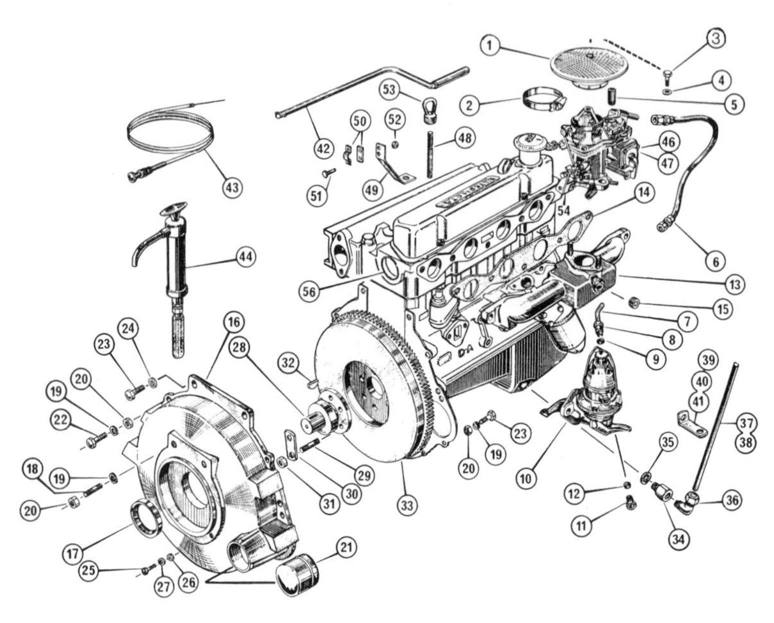

1 Air cleaner
2 Retaining clip
3 Setscrew
4 Washer
5 Pipe carb. vent ($\frac{5}{16}$" O/D × 1$\frac{3}{4}$" O/A)
6 Fuel pipe (flexible) c/w nuts/olives
6A Connector carb
7 Connector pipe (fuel pump)
8 Nut
9 Olive
10 Fuel pump
Hand-primer
11 Nut input side of fuel pump
12 Olive $\frac{5}{16}$"
13 Induction manifold
14 Joint
15 $\frac{3}{8}$" BSP plug (blanking water chamber)
16 Adaptor (gearbox)
17 Starter adaptor
Oil seal for adaptor
18 Stud G.B. retaining ($\frac{3}{8}$" BSF)
19 Spring washer $\frac{3}{8}$"
20 Nut $\frac{3}{8}$"
21 Starter cap
22 Bolts ($\frac{3}{8}$" BSF × 1$\frac{1}{2}$")
23 Retaining bolt coil bracket ($\frac{3}{8}$" W × 1$\frac{1}{8}$")
24 Shake proof washer
25 Bolt matchplate ($\frac{1}{4}$"W × $\frac{3}{4}$")
26 Nuts $\frac{1}{4}$" W
27 Spring washers
28 Stub shaft
29 Studs
30 Clip stub shaft
31 Nuts $\frac{3}{8}$" UNF
32 Stub shaft key
33 Flywheel
34 Sump adaptor nut
35 Fibre washer
36 Elbow union
37 Dipstick tube $\frac{3}{8}$" O/D = 10$\frac{1}{2}$" (Sea Wolf)
38 Dipstick tube $\frac{3}{8}$" O/D = 12$\frac{1}{16}$" (Sea Tiger)
39 Clip/tube (Sea Wolf)
40 Clip/tube (Sea Tiger)
41 Clip/tube (Z-Drive 20 ctrs./spacer block)
42 Cranking handle
43 Choke control. Designate 4′ 6″ or 8′ 6″
44 Sump pump
44A Union
44B Pipe (sump pump)
44C Tube (sump conn.)
45 Tool kit
46 Carburettor (Sea Wolf)
47 Carburettor (Sea Tiger)
47A Carburettor (Sea Leopard)
48 Stud cylinder head Morse attachment
49 Throttle bkt. (engine) }
50 Clip throttle cable } Fitted to rocker
51 Bolt throttle cable } cover post. Jan. '72
52 Nut throttle cable }
53 Lifting eye
54 Throttle lever
55 Cable terminal kit (Morse)
56 Core plugs (stainless)
56A Aft. plug
56B Bkt. Morse

Plate 1.1 *Watermota Petrol Engine*

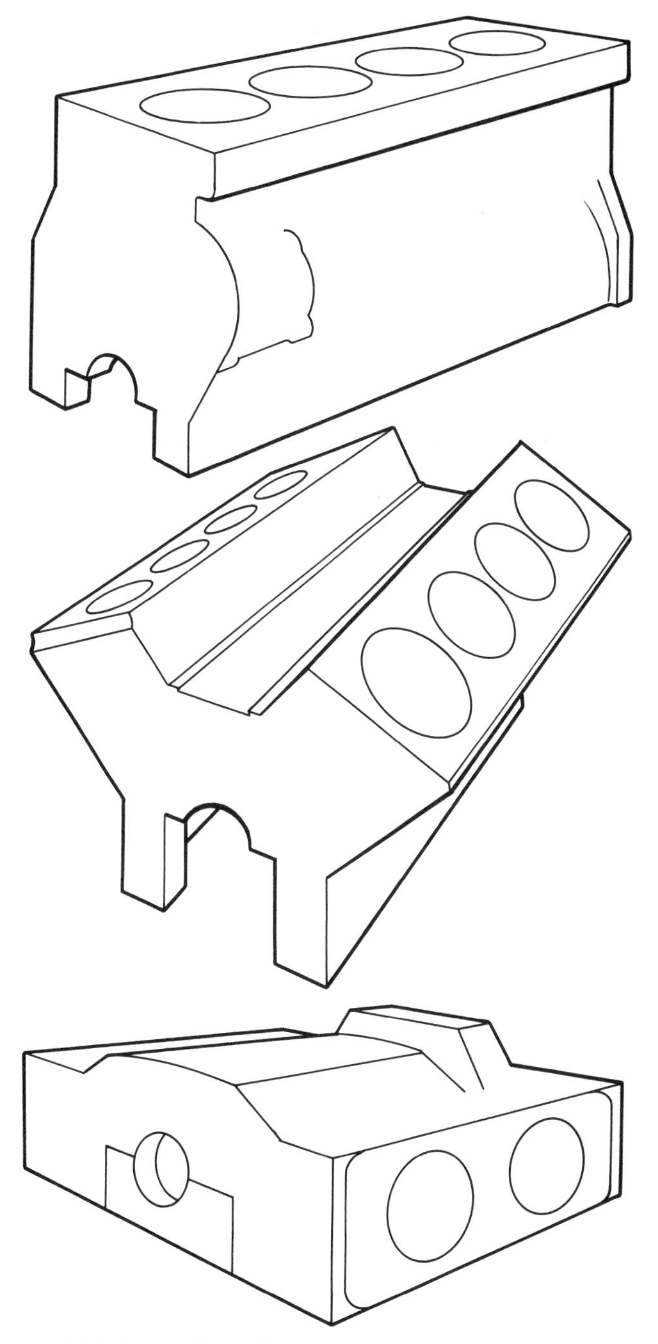

Figure 1.1 *Engine configurations*

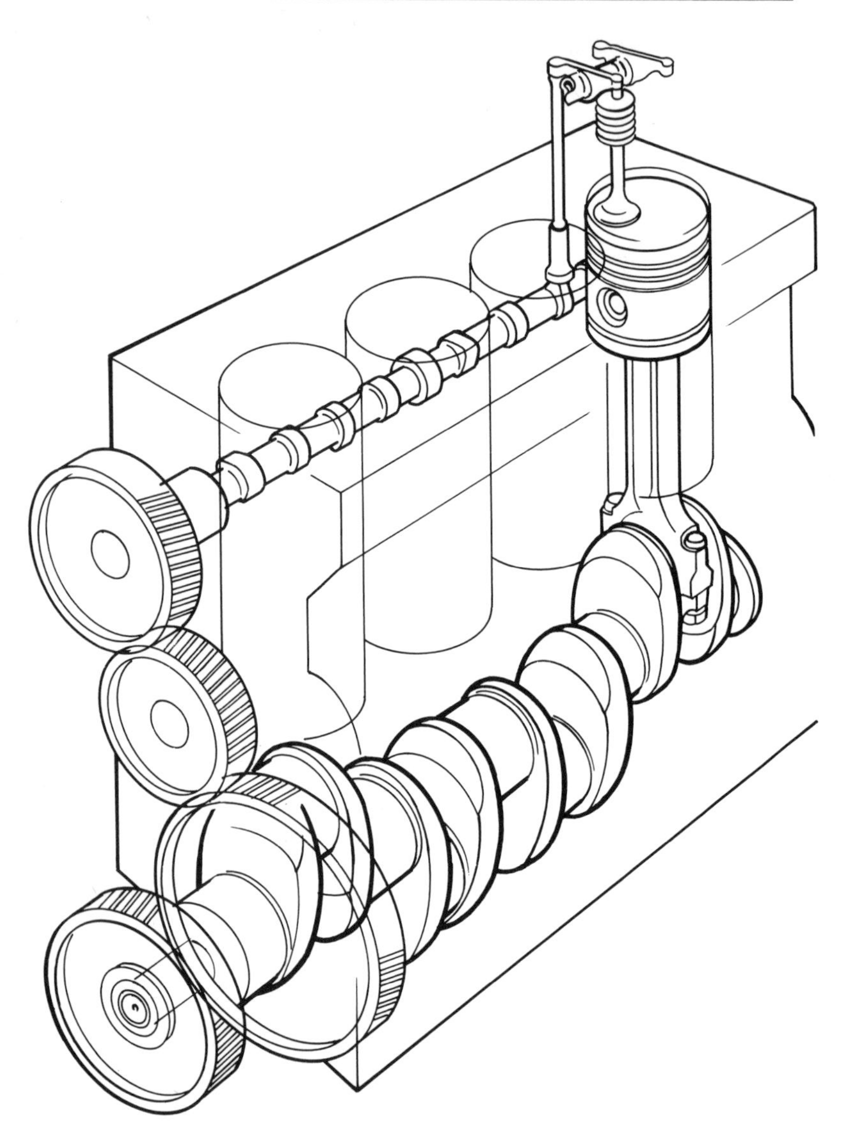

Figure 1.2 *Engine mechanism. The main components of a four-cylinder, in-line overhead-valve engine are indicated.*

Induction

As the piston descends a mixture of petrol vapour and air is drawn through the open inlet valve into the cylinder.

Compression

The piston rises, the inlet valve closes and the mixture is compressed.

Firing

As the piston reaches the top of the cylinder (or fractionally before) the sparking-plug ignites the mixture so that the pressure of the hot gases produced drives the piston downwards to the bottom of the cylinder.

Exhaust

The piston rises, the exhaust valve opens and the spent gases are forced out through the exhaust port. (Figure 1.3)

The process is repeated continuously, the cylinders operating in sequence – each one firing once in two crankshaft revolutions. There are therefore two power pulses for each revolution in the four-cylinder engine. The jerky effect of the firing strokes is reduced by the flywheel, which steadies the motion of the crankshaft by means of its weight.

Numbering the cylinders from the front of the engine, the firing order is usually 1, 3, 4, 2 or 1, 2, 4, 3, either of which gives a more even distribution of firing pressure on the crankshaft than having the cylinders fire one after the other, i.e. 1, 2, 3, 4 followed by a big jump to number 1.

The inlet and exhaust valves are operated by the camshaft, as shown in Figure 1.2. The angular position of the cams on the shaft causes the tappets to rise and fall at predetermined intervals, so that the valves open and close at the best time for the intake of the mixture and expulsion of exhaust gases. The idea is to get as much petrol/air mixture as possible into the cylinder, and then to seal it in by having a leak-proof valve seating in the cylinder-head. Similarly the exhaust valve must allow as much as possible of the exhaust gases to escape, so that little remains to contaminate the fresh charge of petrol/air mixture.

Figure 1.2 shows the most common valve-in-head (or overhead-valve) layout, but older designs have the valve in the top of the

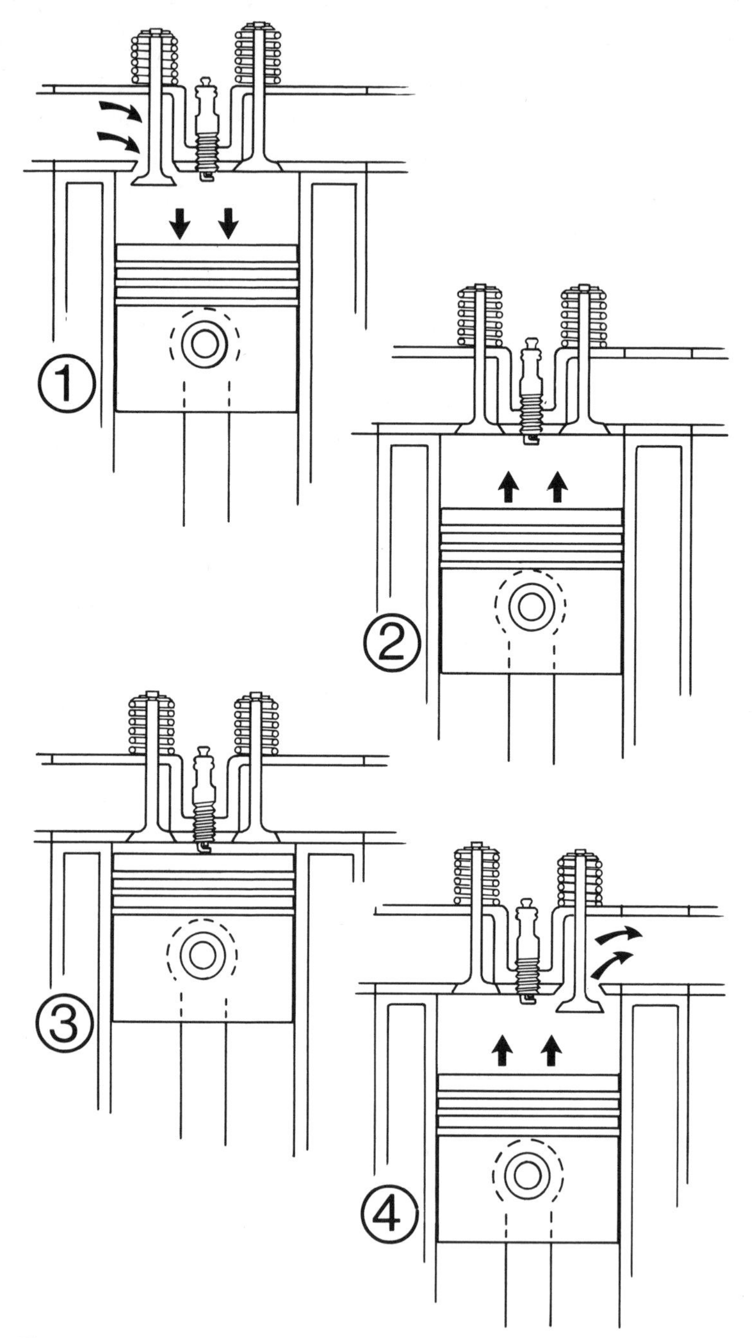

Figure 1.3 *The four-stroke cycle, 1. Induction; 2. Compression; 3. Firing; 4. Exhaust*

cylinder-block – known as the 'side-valve' arrangement. (Figure 1.4)

The camshaft of a four-stroke petrol or diesel engine runs at half engine speed and is generally driven from the crankshaft by gears or chain and sprockets. On some engines the camshaft actuates overhead valves directly, without tappets and rocker shaft. With this arrangement the camshaft is driven from the crankshaft by a chain or by means of a toothed belt and special grooved pulleys. With the camshaft thus located above the valve gear the engine is designated overhead camshaft (OHC). (Figure 1.5)

The piston and connecting rod assembly is shown in Plate 1.2. The pistons are lightweight aluminium components fitted with two or sometimes three rings of cast iron or steel, which press tightly on to the walls of the cylinder to seal the gases above the piston and also keep the lubricating oil below it. Worn rings do not seal properly and hence lubricating oil is burnt and passes out with the exhaust, giving it a characteristic blue colour. The connecting rods have a *small end* with a bush, into which the gudgeon pin of the

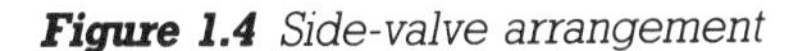

Figure 1.4 *Side-valve arrangement*

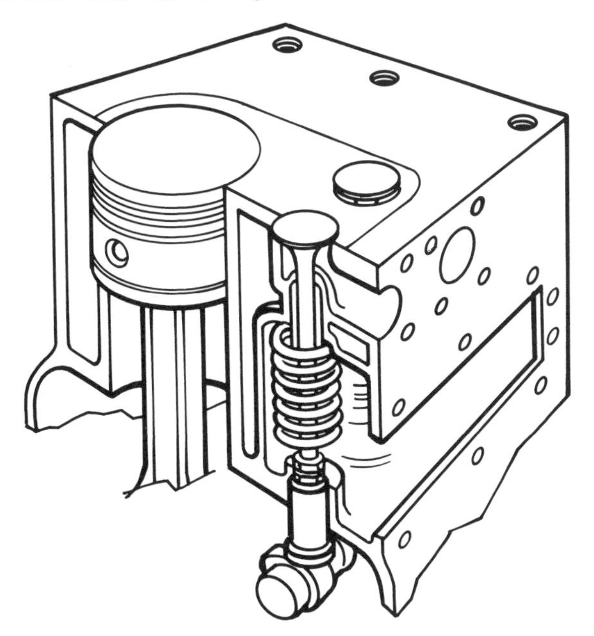

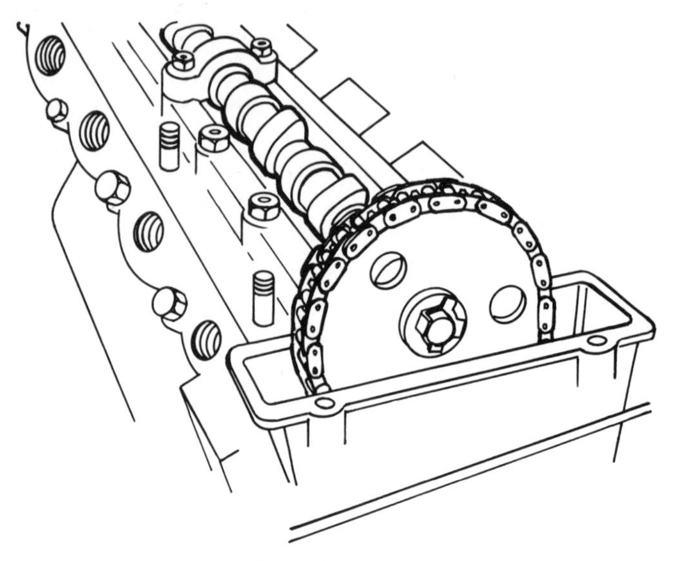

Figure 1.5 *Overhead camshaft. The illustration shows a camshaft driven by sprocket and timing chain enclosed in an oil bath.*

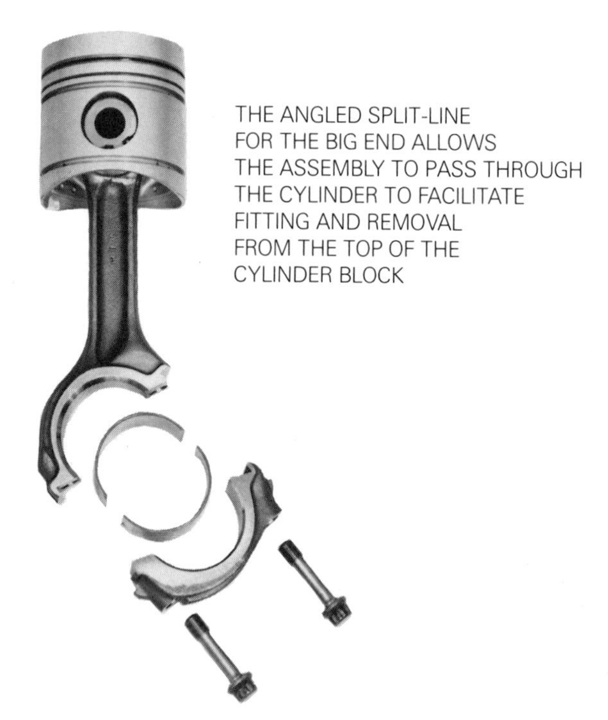

Plate 1.2 *Piston and connecting rod assembly from Perkins engine*

piston is fitted. The *big end*, with its cap and securing bolts, locates the big end bearings which run on the crankshaft *pins*.

The crankshaft is supported by the *main* bearings fitted into the cylinder-block. Drilled holes through the crankshaft direct the pressure oil supply to the main and big end *bearings*.

Induction and Carburettor

The group of pipes connecting the carburettor with the inlet port to each cylinder is called the induction manifold. Air is sucked through the port by the piston on its down stroke when the inlet valve is open. Air rushing through the carburettor on its way to the cylinder passes through the choke of the carburettor and, because the pressure in this area is lower than atmospheric, it draws petrol from the reservoir in the float chamber via a jet into the air stream, where it mixes to form the explosive charge. The choke usually has a reduced section in the centre which speeds up the air flow and increases the suction. Incidentally this choke should not be confused with the air strangler 'choke' which enriches the petrol/air mixture to aid cold starting. (Figure 1.6)

Figure 1.6 *Induction system – petrol engine*

1 Air filter (or flame trap)
2 Air strangler
3 Petrol supply pipe
4 Carburettor
5 Float chamber
6 Float
7 Main jet
8 Idling jet
9 Choke (venturi)
10 Throttle valve
11 Induction manifold
12 Inlet port
13 Inlet valve

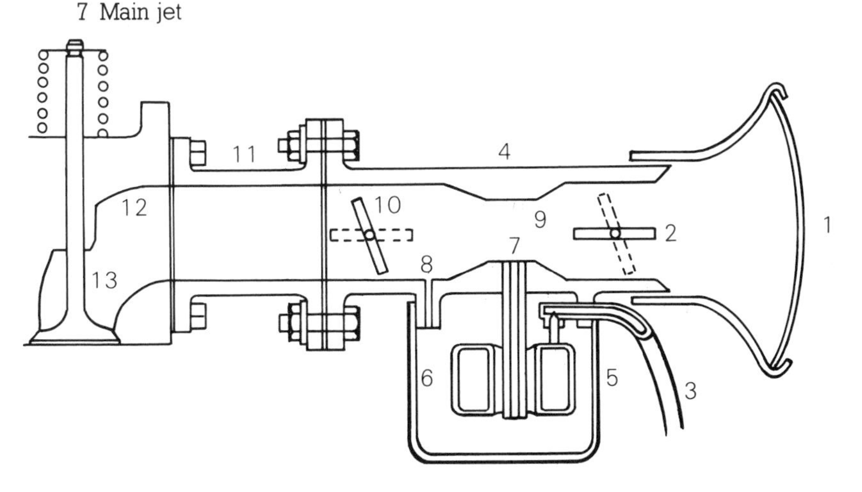

In the simple carburettor illustrated there are main and idling jets which can be adjusted to calibrate the fuel flow. More sophisticated carburettors have several jets which provide for efficient idling, part throttle and full throttle conditions.

Several carburettor models have a variable jet. In the SU type illustrated the jet needle rises and falls with the choke piston, and the varying clearance between the tapered needle and the jet provides the changes in fuel delivery to match the different engine speeds and power required. Jet adjustment is provided by a nut underneath the body which raises and lowers the jet in relation to the needle. On model HD illustrated in Figure 1.7, jet adjustment is provided by means of a screw accessible from above. The cold starting 'choke' moves the jet downwards to enrich the mixture. (Figure 1.7)

Engine speed is varied by the throttle valve which restricts the aperture in the choke tube being almost closed for idling.

Ignition

This system produces a spark in the cylinder when the petrol/air mixture has been compressed (Figure 1.8). The *ballasted* ignition system illustrated in Figure 1.9 improves cold starting by maintaining voltage to the coil when the starter is in operation.

The distributor is a mechanical device which enables the sparking plugs to function at precisely the correct time, which is just before the piston reaches the top of the cylinder. The 'make and break' device in the distributor allows the battery voltage to pass through the primary windings of the coil, and this causes a high voltage to be induced in the secondary windings and then distributed by the rotor arm to the contacts of the leads connecting the plugs with the distributor cap. The 'points' of the contact-breaker open and close by means of four cams – i.e. flat surfaces – at the end of the distributor shaft. The spark occurs as the points open. The condition of the points and the gap between them are the cause of many breakdowns in a petrol engine and they merit careful and regular attention – as do all parts of the ignition system, both mechanical and electrical.

Refinements in the distributor include an 'advance and retard' device operated by flyweights which twist the rotor arm by centrifugal force as the engine runs faster in order to advance the spark, i.e. to start the mixture burning earlier in the cylinder. As the engine slows down the process is reversed. There may also be

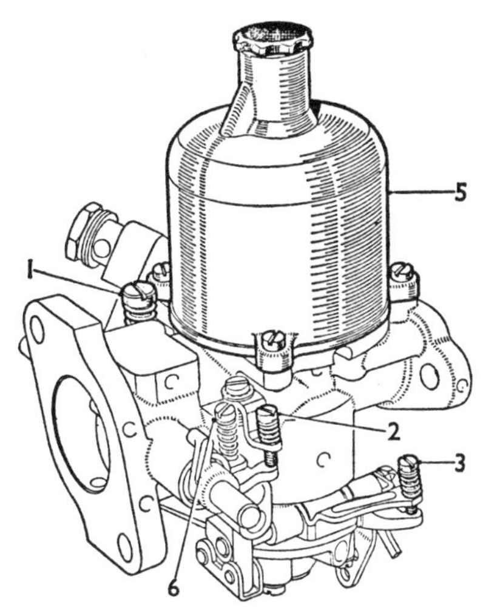

1 Slow-running valve
2 Fast-idle adjusting screw
3 Jet adjusting screw
4 Piston lifting pin (see Figure 4.8)
5 Piston/suction chamber
6 Throttle adjusting screw (when fitted)

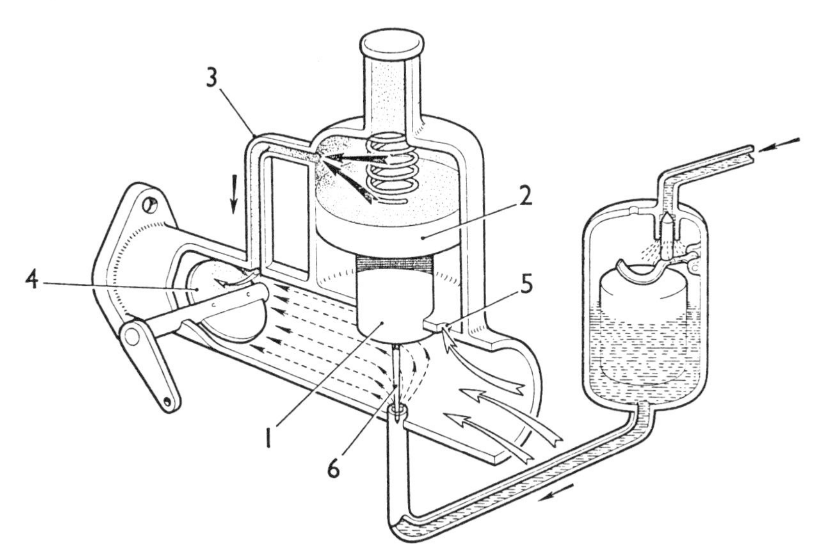

1 Piston
2 Suction disc
3 Drilling through the piston – shown diagrammatically as an external duct communicating the depression caused by the throttle disc to chamber above the suction disc
4 Throttle disc
5 Vent hole to atmosphere
6 Tapered jet needle

Figure 1.7 *Carburettor SU type HD*

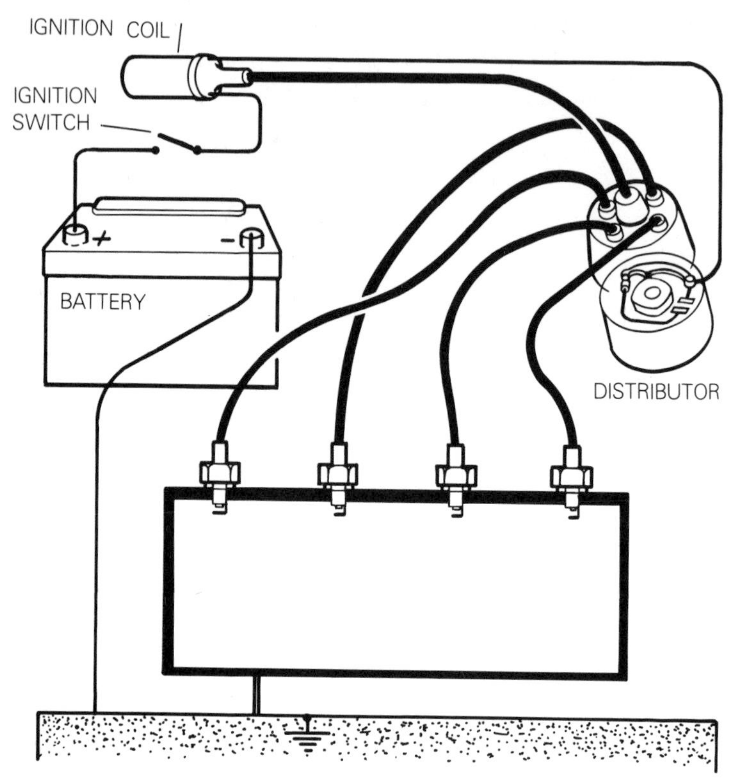

Figure 1.8 *Ignition system*

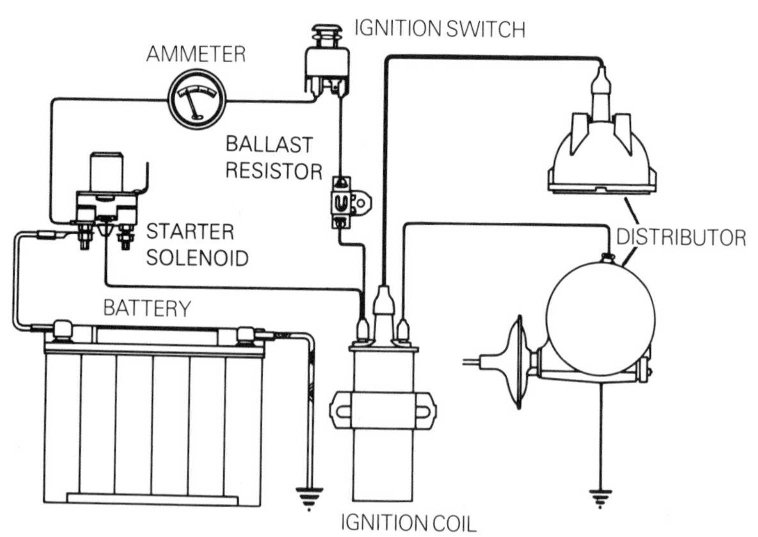

Figure 1.9 *Ballasted ignition system*

another adjustment to the timing which is operated by the load on the engine as sensed by the vacuum in the carburettor. As the load increases the carburettor vacuum is reduced and the ignition timing is retarded. (Figure 1.10)

A *condenser* or *capacitor* located adjacent to the contact-breaker is wired into the primary circuit and has the effect of 'condensing' the current, thus preventing burning of the contact-breaker points.

Figure 1.10 *Distributor assembly*

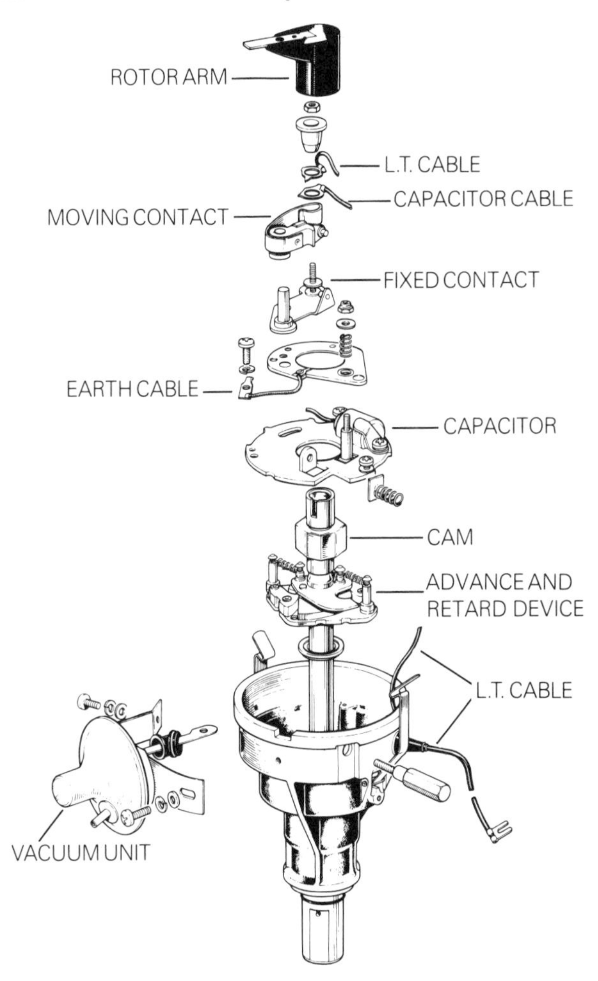

The sparking-plugs are screwed into the cylinder-head from which the high voltage (which may exceed 12,000 volts) is insulated by a porcelain sleeve. The need to change plugs regularly and to keep them in good condition is well publicized for cars and is even more important for boat engines, where an additional hazard is caused by the damp environment. (Figure 1.11)

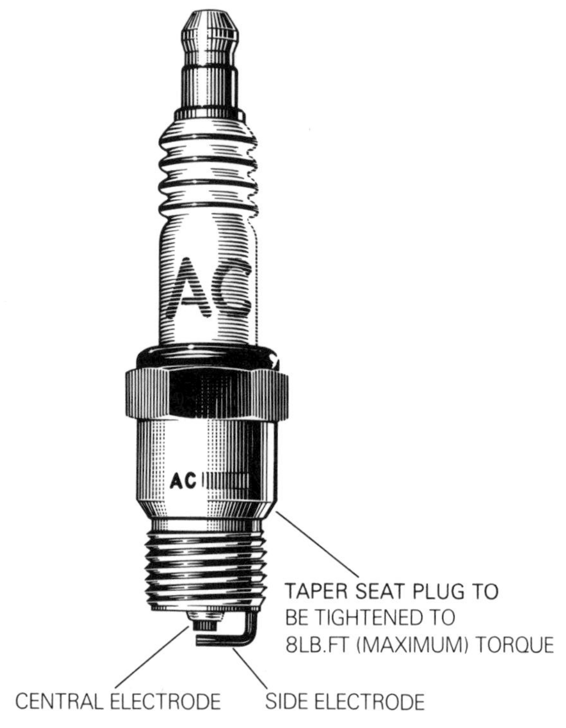

Figure 1.11 *Sparking plug. The model illustrated has a taper seat and does not require a sealing washer.*

Electronic Ignition

This overcomes some of the problems inherent in the ignition system by eliminating the contact breaker and also the condenser. The Bosch TCI-i system, for example, utilizes a pulse-generator, housed in the distributor. A permanent magnet, inductive winding and the core form the stationary enclosed *stator*. A *trigger-wheel* rotates with the distributor shaft via a centrifugal advance and retard mechanism. For a four-cylinder engine there are four projections from the trigger-wheel which pass close to corresponding teeth on the *pole-piece* of the stator, thereby inducing alternating current in the windings of the stator core. The width of the air-gap between the trigger-wheel and pole-piece projections is critical, but because there is no physical contact this does not vary like contact breaker points clearance. (Figure 1.12) Pulses from the distributor pass to a 'trigger box' which rectifies the pulses, controls the dwell angle and stabilizes the voltage.

Figure 1.12 *Electronic ignition system. Bosch TCI-i pulse generator.*

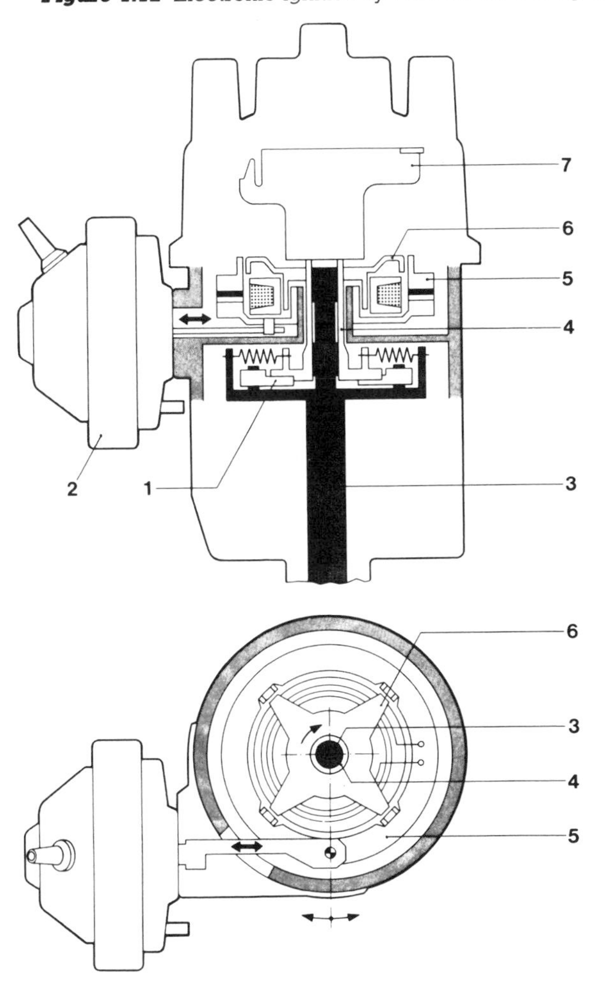

1 Centrifugal advance mechanism
2 Vacuum advance mechanism with vacuum unit
3 Ignition distributor shaft
4 Hollow shaft
5 Pole-piece
6 Trigger-wheel
7 Distributor rotor

As with the conventional ignition system, the coil is usually arranged to operate normally at about 6–7 volts with a 12-volt battery system, but full battery voltage is automatically applied when the starter is used, to facilitate starting.

Exhaust

This system is quite simple, consisting of a pipe or group of pipes known as the exhaust manifold. The outgoing exhaust gas is conducted away from the cylinder head to atmosphere, although the marine engine exhaust may first be mixed with water.

In the marine engine the manifold is usually water jacketed to keep the external surfaces cool and to reduce the risk of fire in the engine compartment. (Figure 1.13)

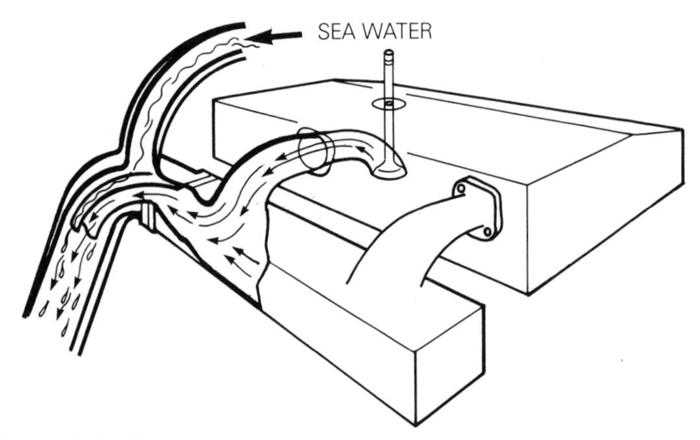

Figure 1.13 *Exhaust system*

Lubrication

An oil supply has to be maintained to all the bearing surfaces in the engine. As well as the crankshaft and camshaft bearings, timing-gears etc., the surfaces of the piston must be lubricated as well as the valves in the cylinder-head. A combination of pressure feed from an oil pump and splash lubrication is used as shown in the diagram. (Figure 1.14)

The oil supply is maintained in the sump, the level of which is checked by means of a dipstick. Oil is drawn from the sump by the

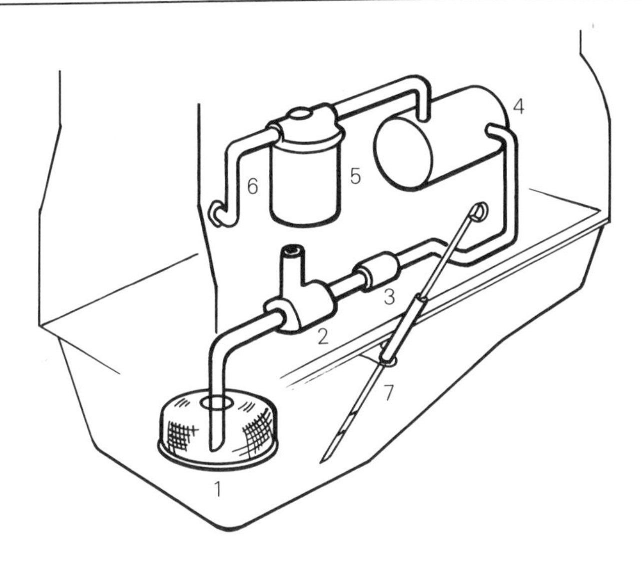

1. Suction pipe and strainer
2. Oil pump
3. Pressure relief valve
4. Oil cooler
5. Oil filter
6. Oil supply to engine
7. Dipstick

Figure 1.14 *Lubrication system*

pump, is forced through the bearings, dripped onto other surfaces and returned to the sump. Incidentally, as well as lubricating, the oil cools the bearing surfaces and in the process it gets hot. In most marine engines, and less commonly in high performance cars, an oil-cooler is used to reduce the temperature before the oil is returned to the sump. Sea or fresh water is circulated through the oil-cooler, passing through tubes around which the oil is pumped.

A filter – generally a paper element cartridge type – is used to remove dirt particles, and a strainer is attached to the end of the suction pipe in the sump.

Cooling

The majority of inboard boat engines are water-cooled by means of sea or river water which is pumped through the cooling circuit and then discharged overboard.

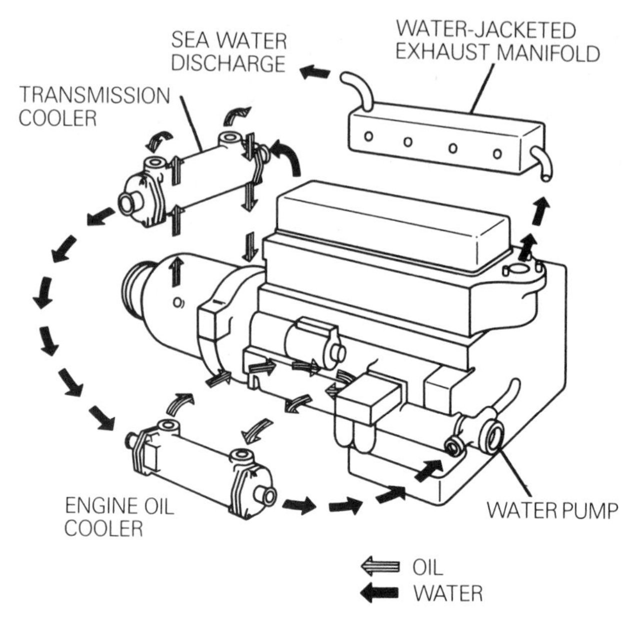

Figure 1.15 *Direct cooling*

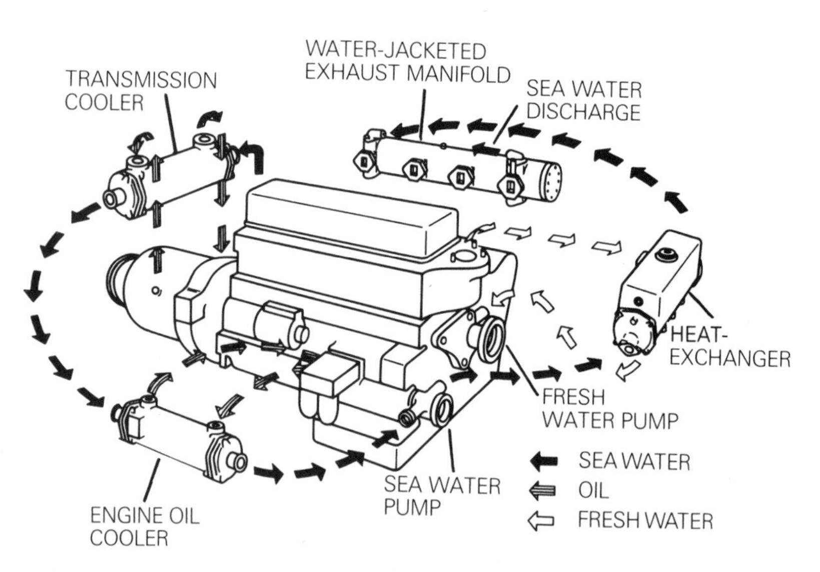

Figure 1.16 *Heat-exchanger cooling*

There are two systems in common use for inboard engines: *Direct Cooling* whereby water is circulated directly through the engine, as the name suggests, and *Heat-Exchanger*, sometimes called indirect cooling, whereby the water goes into a heat-exchanger to cool a secondary circuit of fresh water which is pumped through the engine. The advantage of heat-exchanger cooling is that the often corrosive and gritty sea or river water does not actually pass through the narrow spaces in the cylinder-head to cause possible blockages and corrosion. Both systems are illustrated by Figures 1.15 and 1.16.

Some small engines are air-cooled. Inboard models require ducting to conduct the air flow to and from the engine, and a fan is used to circulate the air. (Figure 1.17)

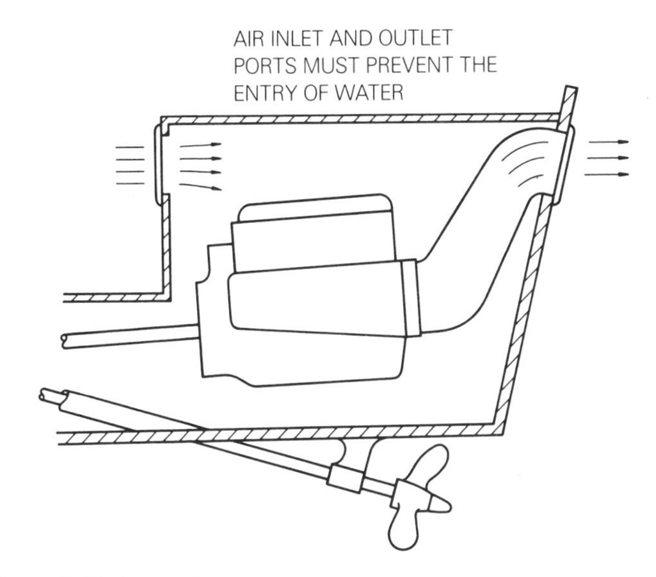

Figure 1.17 *Air cooling*

Two-Stroke Cycle

By arranging the four phases of the combustion cycle so that they occur during one revolution of the crankshaft, more power can be produced from the same size of engine, which then is called a 'two-stroke'. Alternatively the same power can be produced with a smaller, lighter engine, which is one of the reasons for this system being favoured for outboard motors. However, although many ingenious two-stroke designs have been produced there is a reduction of efficiency, mainly caused by the difficulty in getting rid of the exhaust before compressing a fresh charge of air-and-fuel mixture.

As the complete cycle occurs in one rise and fall of the piston, use is made of the space underneath the piston (i.e. the crankcase volume) to store the air/fuel mixture, which is admitted via a 'leaf' or 'reed' type valve in the simple design shown here. This reacts to the pressure in the crankcase, opening to admit air/fuel mixture from the carburettor when there is a depression, and closing when it is pressurized. The piston creates pressure when falling and causes a depression when rising.

It is a little difficult to say where the two-stroke cycle starts, but we can pick up the sequence with the piston at the bottom of the cylinder. The air/fuel mixture was sucked into the crankcase while the piston was rising and will now be under pressure with the valve closed; but when the piston top uncovers the exhaust and inlet ports in opposite sides of the cylinder wall, the mixture will be forced through the intake port and deflected towards the head by the baffle on the top of the piston, thereby helping to clear the remainder of the exhaust gases through the port.

While the piston is rising the valve opens, sucking in a fresh charge of air/fuel mixture. As the piston closes both ports, the mixture is compressed and then ignited by the sparking-plug at, or just before, the top of the stroke. As the piston is forced down on its power stroke by the pressure of the hot gases it uncovers the exhaust and inlet ports, thus preparing the cylinder for the next cycle. (Figure 1.18)

Outboard Motor

The type of two-stroke engine used for outboard motors is purpose-made for marine use, with lightweight aluminium castings where the inboard four-stroke has cast iron. The usual number of cylinders is one, two, four or six, and the crankshaft is positioned in the vertical plane. There are no valves in the cylinder-head; the intake and exhaust operations are carried out via ports in the side of the cylinders and the crankcase, as shown in Figure 1.19.

Two-Stroke Lubrication

The outboard motor two-stroke is essentially a simple engine in

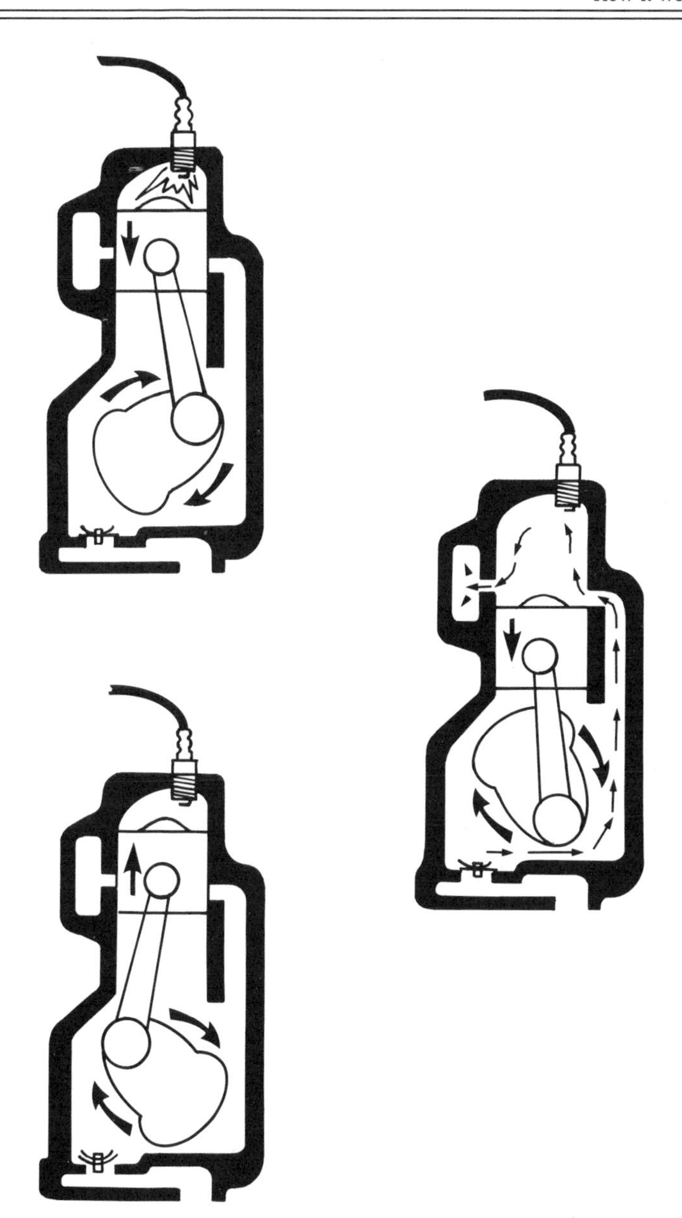

Figure 1.18 *The two-stroke cycle*

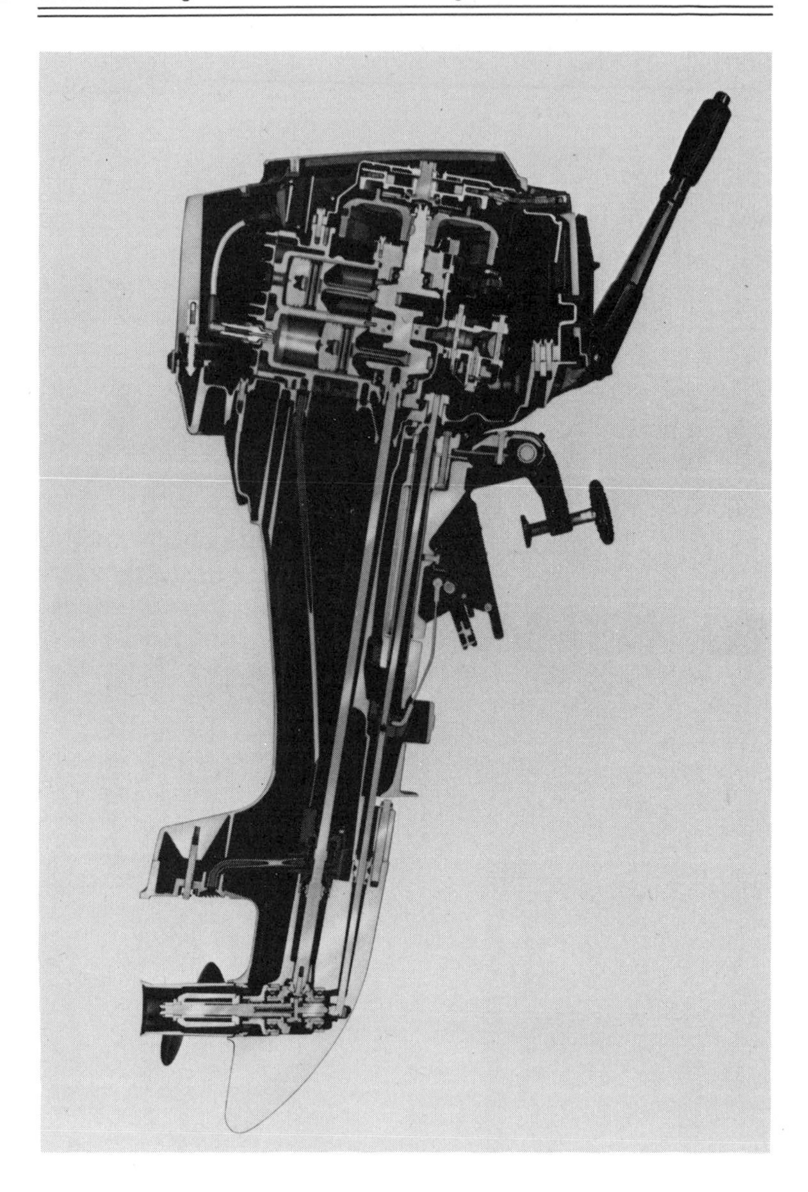

Figure 1.19 *Outboard motor – Mercury Marine*

principle, although the power head of one of the more powerful multi-cylinder models may not appear so. However, the lubrication system is simple: you just mix the required amount of oil with the petrol, and the fact that all the moving parts inside the engine are in contact with the air/fuel mixture ensures that they are thoroughly lubricated. Too much oil in proportion to the petrol will cause plug fouling, and too little may cause an engine seizure.

The transmission gearing of the outboard motor has a separate lubrication system similar to the stern-drive 'leg'.

Diesel

So far we have been mainly concerned with inboard petrol engines, but much of the text applies equally to an inboard diesel or 'compression-ignition' engine, to use a more correct title; Dr Diesel's engine did not in fact operate precisely as the modern 'diesel' does.

The main difference between petrol and diesel engines is that whereas the former draws a mixture of petrol vapour and air into the cylinder, compressing it to about one eighth of its volume and then igniting it with a spark, the diesel compresses air to a much higher pressure and then injects fuel oil, which mixes with the air and ignites without the need for a spark. (Figure 1.20)

Figure 1.20 *Diesel air induction system*

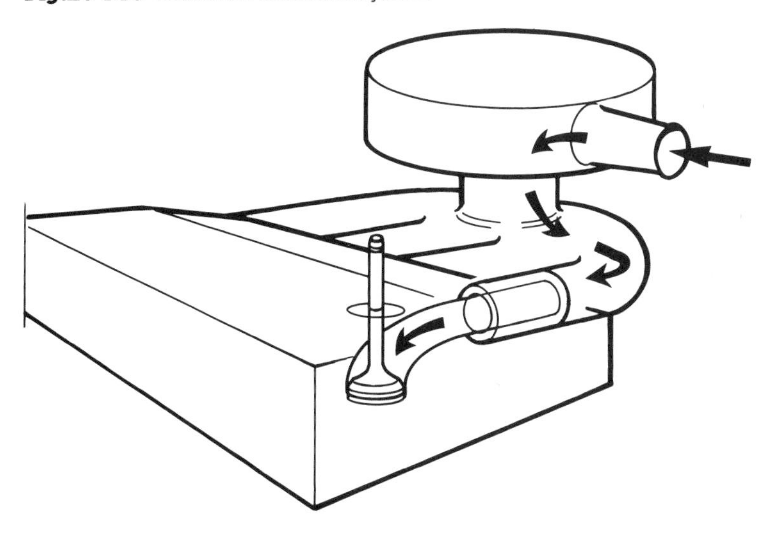

The explosion in the diesel cylinder creates higher pressures, making it necessary for the engine to be stronger than a petrol motor of the same bore and stroke. Diesels therefore tend to be heavier than their petrol counterparts. However, they generally run more reliably and for longer periods before needing an overhaul.

Figure 1.21 shows the main components in a diesel engine's fuel-injection system.

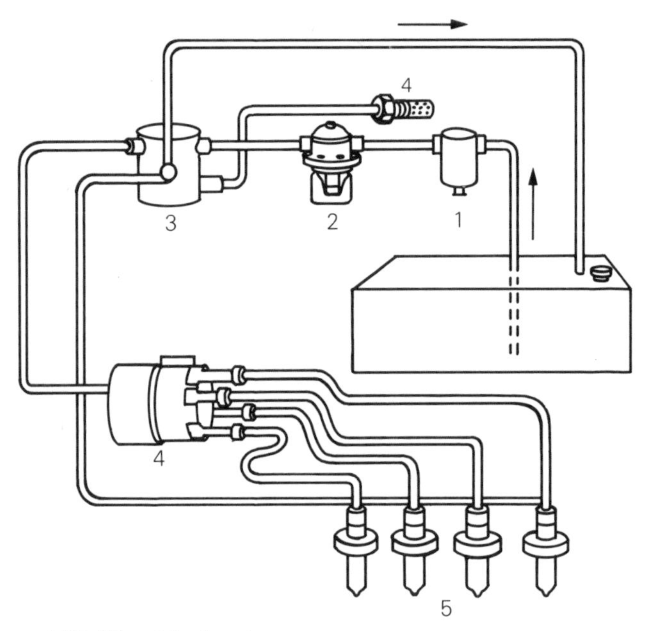

Figure 1.21 *Diesel fuel system*

1. Fuel pre-filter water-trap
2. Fuel lift pump
3. Fuel filter
4. Fuel injection pump
5. Fuel injectors

The heart of the fuel-injection equipment, which incidentally is referred to by engineers as the 'F.I.E.', is the fuel-injection pump. This is made to extremely close 'tolerances', i.e. limits of accuracy, in order to prevent fuel leakage and also to meter the minute quantities of fuel which are accurately injected for each power stroke. The timing of these injections is just as important as the ignition timing of a petrol engine, and a sophisticated fuel-injection pump such as the rotary pump shown in Figure 1.22 has a means of varying the timing according to the varying needs of the engine. Maximum efficiency is thus assured and hence economical running.

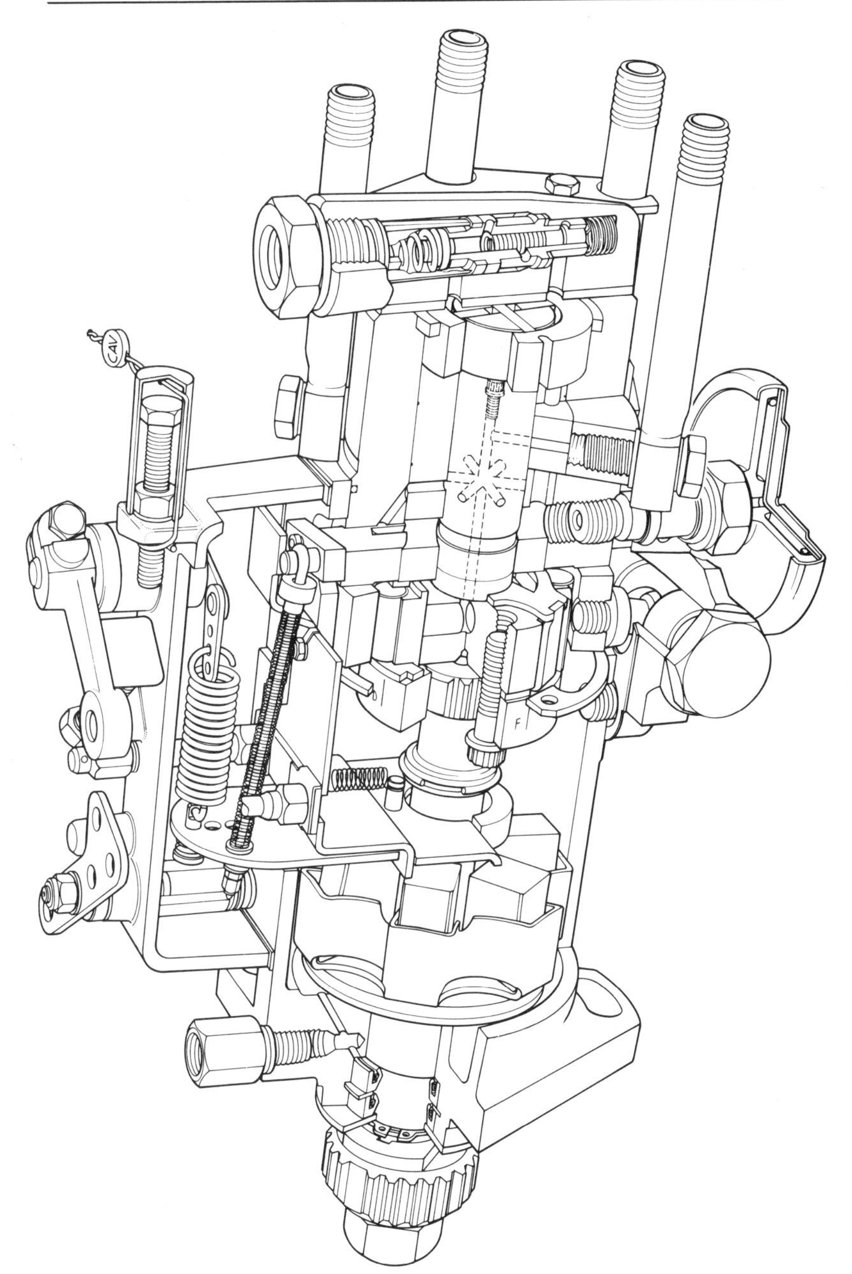

Figure 1.22 *Rotary fuel injection pump. Lucas CAV type DPA incorporating a mechanical governor.*

Like the distributor of a petrol engine, the fuel-injection pump runs at half the crankshaft speed and is driven by a timing gear or sprocket – occasionally from the camshaft. A governor controls the engine speed and works in conjunction with the throttle lever on the outside of the pump, to which is attached the cable or linkage from the helmsman's position. A 'stop control' lever on the pump shuts off the fuel supply and is also operated from the wheel-house – by cable or electrically.

Fuel is squirted into the cylinders at high pressure via injectors – also known as nozzles. These contain a spring-loaded valve which lifts off its seat, sprays the metered quantity of fuel from the pump into the cylinder and then snaps back onto the seat, ready for the next injection.

Two models of injector are shown in Figures 1.23 and 1.24; various types and sizes are used, with one or several holes according to the needs of the engine.

Figure 1.23 *Multi-hole injector. For use with direct-injection diesel engines.*

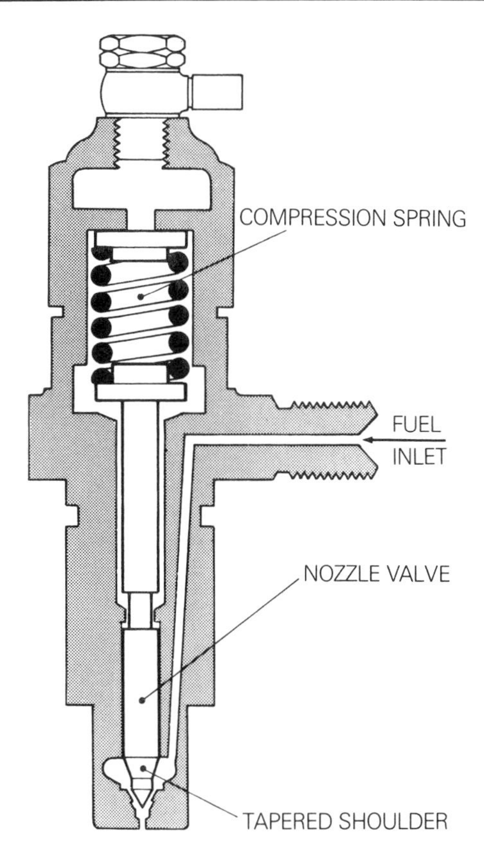

Figure 1.24 *CAV Pintle type injector. Used for indirect-injection engines.*

Four-Stroke or Two?

Like inboard petrol engines, most makes of diesel boat engine operate on the four-stroke principle. The four phases of induction, compression, power and exhaust stroke are the same for both diesel and petrol, with the exception of the injection of fuel in the diesel; and in fact the more advanced petrol engines inject petrol in the same way, so that they are very similar to the diesel engine as regards the operating cycle.

With two-stroke diesel engines, however, the similarity with two-stroke petrol motors is not so great. The diesel does not make use of its crankcase for admission of the 'charge', and a means of 'scavenging' the cylinder is found necessary; this function is carried out by an air-pump or 'blower' driven from the engine. There are different types of scavenging system and various designs of two-stroke engines with or without valves in the cylinder-head. In the Detroit Diesel engine illustrated in Figure 1.25 four exhaust valves for each cylinder are operated by the camshaft. The air is admitted via a ring of ports passing through the cylinder wall pressurized by the blower.

Figure 1.25 *Two-stroke diesel engine. Detroit Diesel Allison model 8V–71.*

Direct or Indirect Injection?

When the diesel combustion chamber is in the piston as shown in Figure 1.26, the system is said to be 'direct injection'. There are many variations in the shape of the combustion chamber, but this arrangement generally gives the best fuel economy and easiest starting.

Engines required to run at higher speeds are usually *indirect injection* as in Figure 1.27 with the combustion chamber in the

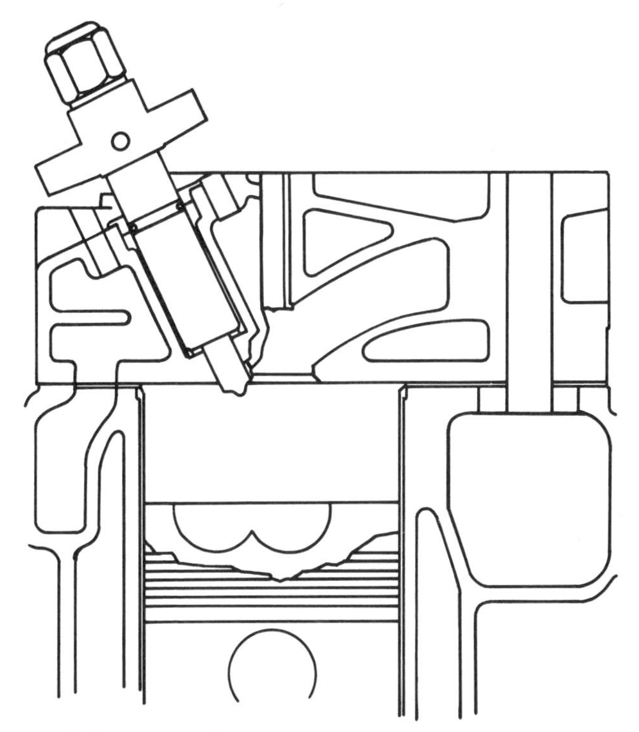

Figure 1.26 *Direct-injection system*

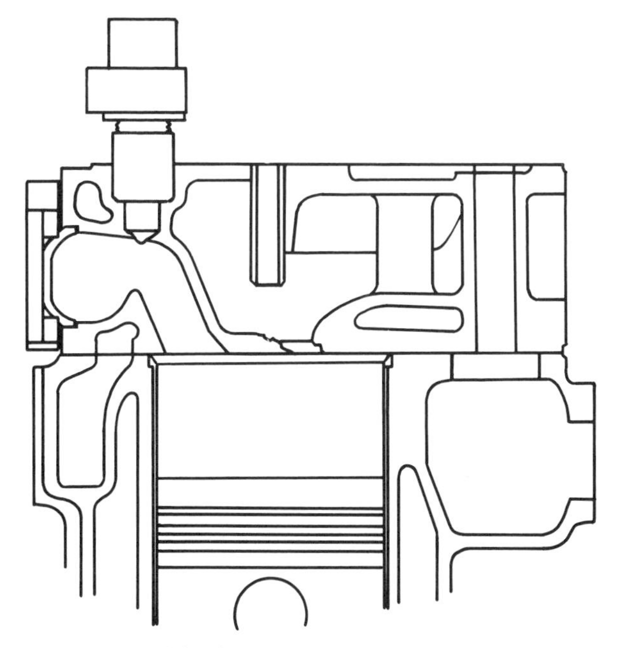

Figure 1.27 *Indirect-injection system*

cylinder-head. These engines usually pay a small penalty of lower efficiency than direct injection with slightly inferior fuel consumption and with the need for a starting aid at higher air temperatures.

Power Measurement

Power output for petrol or diesel engines is usually expressed in horsepower, and when this is measured on the test bed with the engine driving a *test brake*, the power is defined as brake horse power (bhp) In fact it is the engine *torque* that is measured by the test brake and then related to the crankshaft speed in the formula:

$$\text{bhp} = \frac{\text{torque (lbf ft)} \times \text{engine rev/min}}{5252}$$

Metric horsepower is a slightly smaller unit defined as Pferdestärke (PS) or cheval vapeur (ch or CV).

1 bhp = 1·01387 PS, ch or CV

Engine power is also expressed in Kilowatts

1 bhp = 0·7457 Kw

Atmospheric temperature and pressure affect the weight of air used by an engine and slightly affect the power output. Various authorities have defined standard atmospheric conditions at which the engine power should be quoted with formulae for adjusting the power measured on the test bed to what it would be at the standard conditions. British Standard BS Au 141a:1971 is an automotive standard commonly used for diesel engines in boats with standard air temperature 20°C (68°F) and total barometric pressure 760 mm. (29.92 in.). German boat engines frequently use DIN standards and SAE publish standards in the USA.

Turbo-chargers

The power output of petrol and diesel engines is limited by the amount of air that can be induced into the cylinders. There is no difficulty in increasing the amount of fuel, but this must be accompanied by sufficient air to achieve satisfactory combustion.

A turbo-charger can boost the weight of air taken into the engine by about 50 per cent, with a corresponding increase in power. A small centrifugal compressor at one end of the turbo-charger shaft is rotated at very high speed: over 80,000 revs per minute are achieved in small turbo-chargers. The driving force is provided by a turbine wheel attached to the opposite end of the shaft which is driven by the exhaust gas stream. (Figure 1.28)

Figure 1.28 *Turbo-charger. Cut-away Holset model showing exhaust-driven turbine on left coupled to the air compressor. The rectangular flange beneath the turbo-charger would be bolted to the engine manifold.*

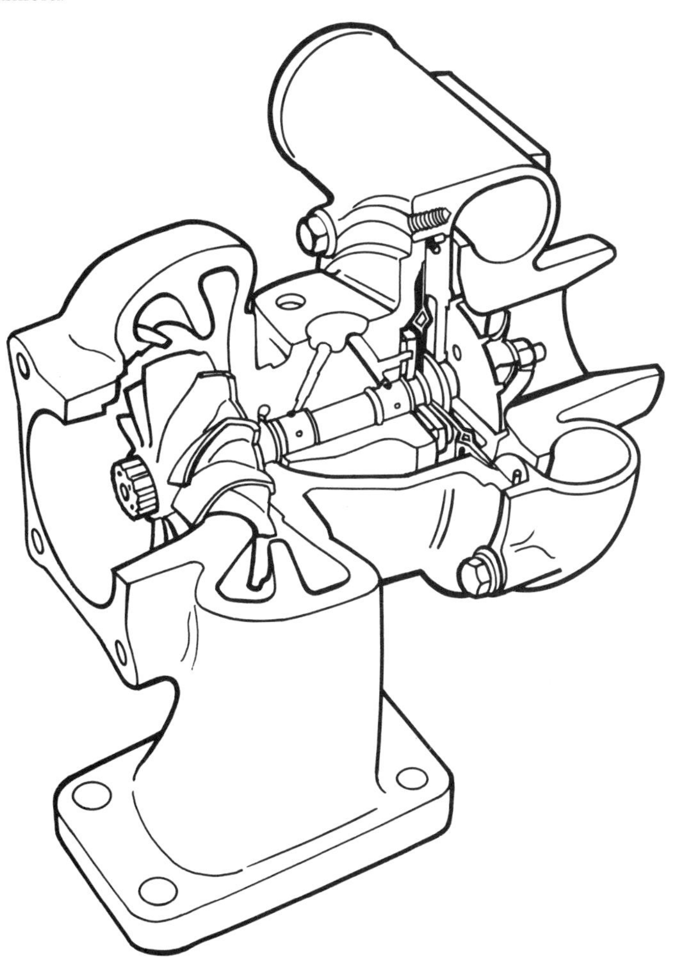

Air-Charge Cooling

A considerable further increase in power for turbo-charged engines can be provided by the addition of an air-charge cooler, sometimes described as an *intercooler.* This counteracts the effect of the turbo-charger compressor which heats the air and in doing so loses some of the advantage gained because the weight of air introduced into the cylinders is reduced as its temperature increases. By the interposition of a cooler between the compressor outlet and the inlet manifold, the air is cooled and a greater weight can be introduced.

With the very large coolers used in conjunction with powerful turbo-chargers on racing engines, both petrol and diesel, three or four times the naturally-aspirated power can be developed, although a ratio of two to one is seldom exceeded for cruiser engines or commercial duty. The upper limit of power is set by the ability of the crankshaft, connecting rods, bearings etc. to withstand the tremendous stresses caused by the high cylinder pressures. Reducing the temperature of the compressed air intake is beneficial to the life and reliability of the engine, because the blast of cool air on each compression stroke helps to maintain the exhaust valves, piston crown and piston rings below critical temperatures.

In the air-charge cooler illustrated in Figure 1.29, air from the

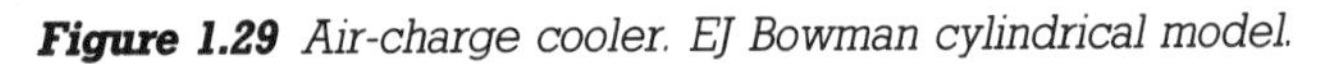

Figure 1.29 *Air-charge cooler. EJ Bowman cylindrical model.*

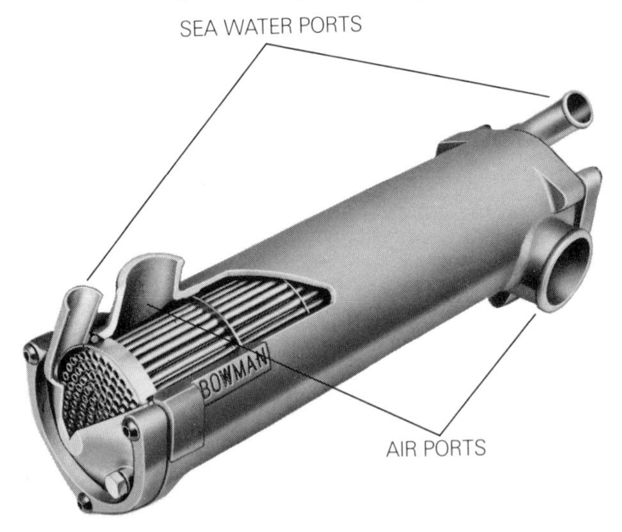

compressor passes through baffled compartments in the body which are cooled by sea water pumped through the stack of tubes.

To save space and reduce pipework on Range 4 engines, Perkins combine the charge cooler with the lubricating oil and transmission coolers in the 'EGA' design shown in Figure 1.30. Another way of saving space is to locate the charge cooler tube stack in the air inlet manifolds.

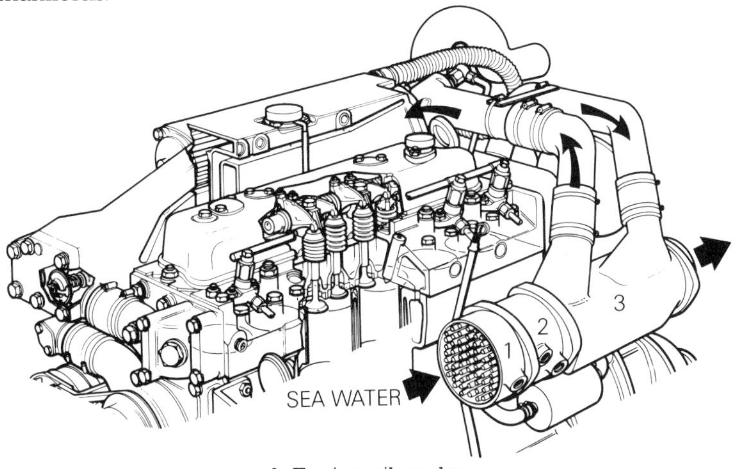

1 Engine oil cooler
2 Gearbox oil cooler
3 Air-charge cooler

Figure 1.30 *Air-charge cooler (EGA type)*

Cold Starting

Petrol and diesel engines do not operate efficiently until working temperature is reached in the combustion area. When an engine has been run for a time the heat produced by combustion will maintain the required operating temperature and the water cooling-jackets insulate this area, carrying away excessive heat. Air-cooled engines achieve the same effect by a blast of air over the cooling fins, although they do not have the insulating effect of a water-jacket. When starting from cold, arrangements have to be made to compensate for the effect of the cold surfaces at the top of the cylinder, in both petrol and diesel engines.

Petrol Engine

The rich mixture needed for starting a petrol engine increases the petrol-to-air ratio to compensate for the loss by condensation of petrol which occurs on the cold surfaces of the inlet manifold and the cylinder walls, cylinder-head and pistons. The ideal air-to-fuel ratio in the area of the spark-plug for easy ignition of the spark is about 14:1, and this is achieved by choking the air inlet, i.e. operating the strangler valve. As the engine warms up and condensation is eliminated, the choke is released. Leaving the choke 'out' too long causes the engine to stall because of too much fuel in the mixture and because of wet plugs, which have to be dried out before the engine will run.

Diesel Engine

With the diesel system of injecting fuel into a cylinder heated to a high temperature by compressing the air, the metal surfaces do not become cold enough to make the fuel condense, and so the problem associated with the petrol engine does not arise. However, at low air inlet temperatures it is necessary to pre-heat the air before starting, because some of the heat produced by the compression stroke is lost to the cylinder walls, piston and cylinder-head, etc. When the engine starts, these surfaces retain sufficient heat to continue running. The diesel engine will produce full power sooner than a petrol engine because the combustion surfaces warm up faster.

Pre-heating of the inlet air for a diesel engine can be achieved in several ways, as illustrated in Figure 1.31. Heater plugs are inserted in the combustion area above each cylinder and are switched on for a specific period prior to operating the starter motor.

Fuel-burning devices such as the Lucas Thermostart introduce fuel to an electrically heated element. These devices do not require as much electrical energy as the heater plug because current is required only to ignite the fuel. The Thermostart is located in the inlet manifold and a single unit is usually adequate.

Ether is used to start diesel engines down to very low temperatures. With thin lubricating oil, about SAE5 viscosity, a start can be obtained below −20°C (−4°F) with a suitable battery and starter motor, particularly with direct injection engines. An aerosol spray can be used to direct ether into the air intake, but indiscriminate use can damage the engine by excessive cylinder pressure, so that an engine-feed device with a nozzle to dispense a limited amount of ether is a safer proposition.

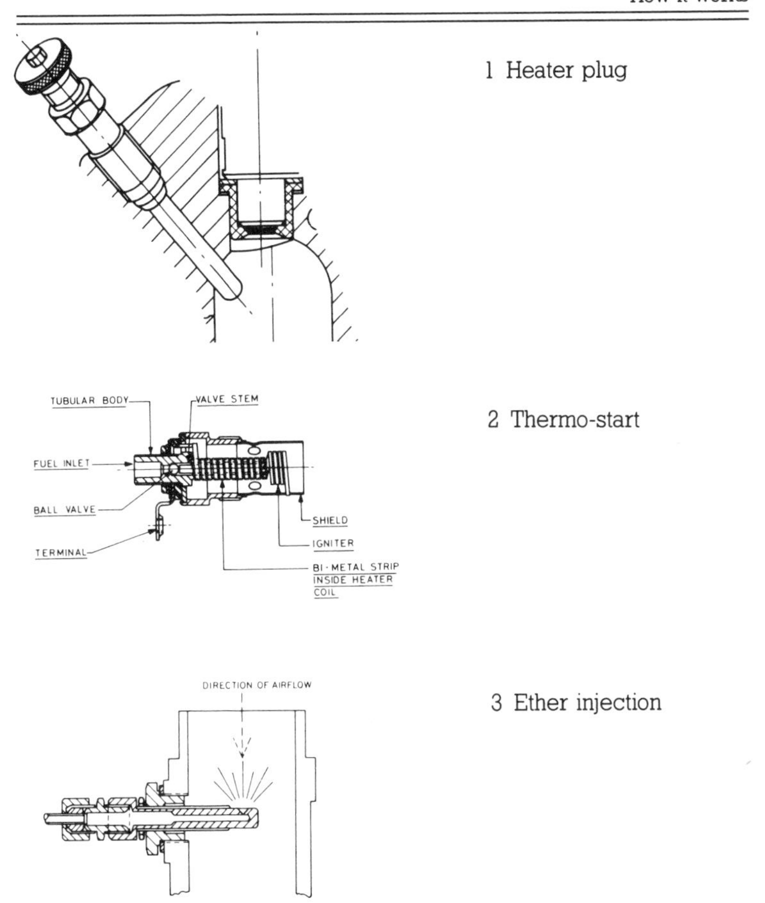

Figure 1.31 *Cold starting aids for diesel engines*

Electrics

Simple, small diesel engines with hand-starting gear as used in small launches or fishing-boats may have no electrical equipment.

In fact, once a diesel engine is running, the electrical equipment, i.e. starter, alternator, regulator, battery, wiring and instrumentation, can be dispensed with – at least until you need to start up again. The petrol engine, of course, needs an efficient battery and alternator to keep the ignition circuit operational.

However, whether petrol- or diesel-fuelled, electrical equipment for boats needs to be of a superior standard to that fitted to cars and other road vehicles because the damp, salt-laden environment will attack the exposed parts of the alternator and starter, especially when the boat is not in use. Non-corrosive materials, better-quality wiring, waterproof connections and a generally higher standard of components have been found necessary to ensure reliability in sea-going conditions.

Starter Motor

This is a small battery-operated electric motor which rotates at about ten times engine cranking speed, its pinion meshing with the flywheel gear ring.

There are two main types of starter, the differences being the method of engaging and disengaging the drive between the pinion and flywheel gear. The 'Bendix' drive is used for many petrol engines and the smaller diesels. As the illustration shows, there is a spiral groove in the shaft, and the pinion engages with this. When the starter motor is operated the shaft accelerates very rapidly, but because the heavy pinion, held away from the flywheel by a light spring, takes a fraction of a second to be accelerated to the starter speed, it moves along the spiral and goes into mesh with the gear ring. When the flywheel accelerates as the engine fires and picks up speed, the pinion is thrown out of mesh and the spring holds it away from the flywheel until the next start. A short heavy spring at the end of the shaft cushions the impact of the pinion as it shoots along the spiral groove. (Plate 1.3)

The positive-engagement type starter does not rely on the automatic engagement of the pinion, which can become unreliable after frequent use – especially in low temperature conditions when the pinion can slip out of mesh before the engine picks up speed. The positive-engagement starter uses a solenoid linked to the pinion, so that this is pushed into mesh with the flywheel gear ring before the shaft starts to turn. This type of motor is easily recognized by the solenoid, which is generally mounted on the starter casing. Also, there are splines on the pinion shaft instead of a spiral groove. The starter has an overrun clutch, which ensures that the running engine cannot drive the starter with its pinion locked in mesh with the flywheel gear. (Plate 1.4)

Alternator

Almost all modern engines use an alternator instead of a dynamo. Although there is an added complication with an alternator since its

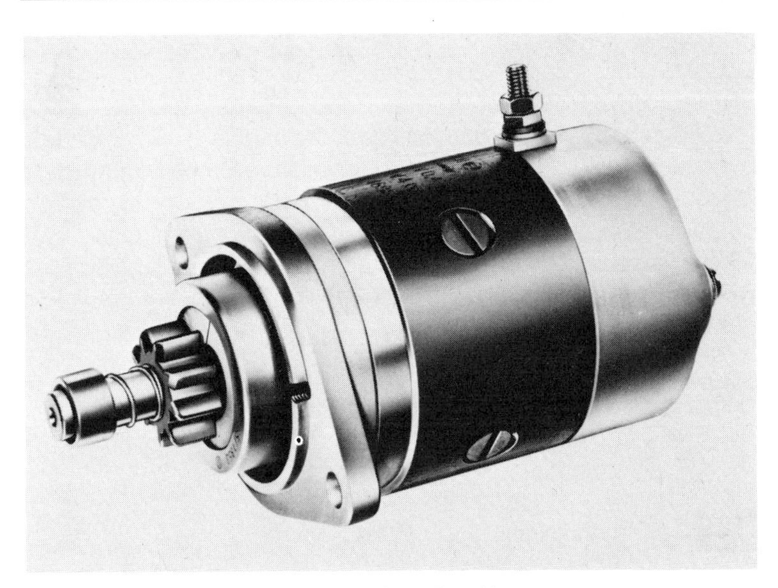

Plate 1.3 *'Bendix' Type Starter (Robert Bosch)*

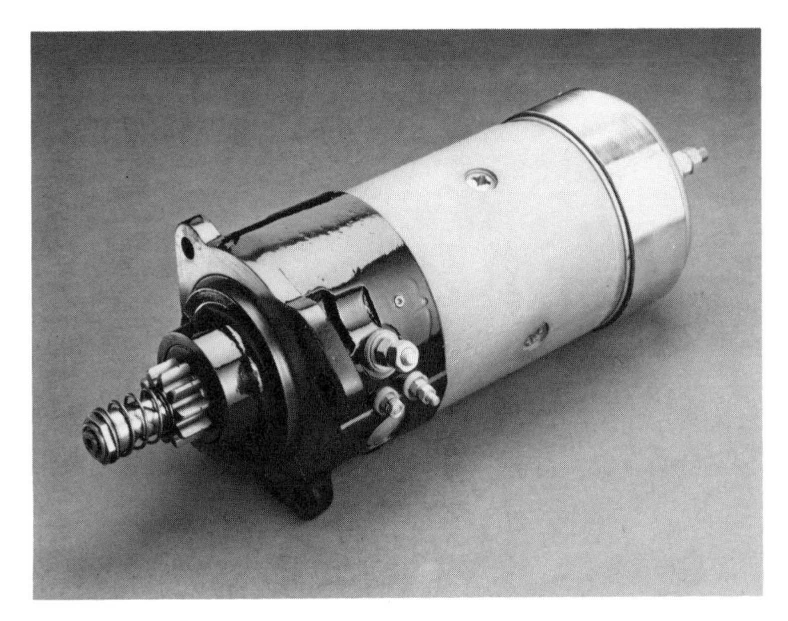

Plate 1.4 *Pre-engaged starter (CAV)*

alternating current has to be rectified, i.e. converted to direct current before it can be used to charge the battery, its efficiency is greater than a dynamo of equivalent size because it runs faster, and it will generally deliver current when the engine is idling, while a dynamo has a cut-in speed several hundred revolutions per minute higher. The increasing need for more current to supply shipboard domestic equipment as well as navigational aids, more lighting, etc., means that engines will now be fitted with 50 to 70 amp alternators whereas twenty years ago 15 to 20 amps from the generator would generally have been sufficient.

Figure 1.32 *Alternator. Motorola type 9AR/9AL.*

A Rear cover
B Clip
C Diode bridge
D Field connector
E Brush holder
F Bracket
G Quadrant
H Rear housing
I 'O' ring
J Stator
K Rotor
L Spacer for rotor
M Rear bearing
N Front bearing
O Front bearing mounting plate
P Front housing
Q Through-bolts
R Fan
S Pulley
T Washer
U Pulley fixings
V1 Clamp
V2 Terminal, earth
W Terminal, D+
X Capacitor
Y Cover
Z Screws

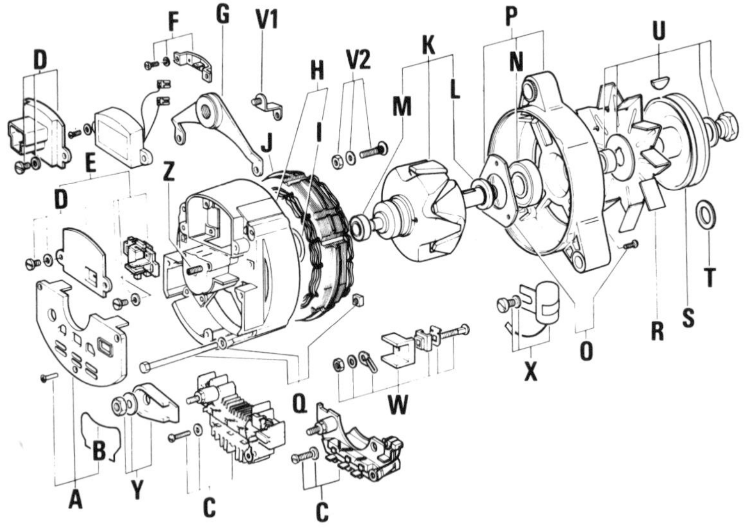

Although alternators have become commonplace and all auto electricians are familiar with their servicing, they are more vulnerable to abuse than a dynamo. The conversion of a.c. to d.c. is carried out by diodes, which also prevent the battery current from reaching the alternator when the engine is not running. If you accidentally disconnect the battery while the alternator is being charged, or connect the positive and negative cables to the wrong terminals, the diodes will fail and a reconditioned alternator or new set of diodes is required.

A regulator is used to control the voltage output from the alternator. For a nominal 12-volt machine the voltage is about 14. The latest alternators have a built-in regulator.

Batteries

Lead-acid or alkaline batteries are used in boats, and the system voltage is generally 12 or 24. The capacity of the battery obviously has to be adequate to start the engine and operate the auxiliary current requirement. On large craft an independent auxiliary engine-driven generator is often used, to avoid the need to run the main engines when at the moorings, just to keep the batteries charged. Very small, portable generators are available and are extremely useful if the alternator fails or you spend days in harbour waiting for the weather to improve without a mains electrical supply.

It is good practice to have separate batteries for engine-starting and 'general services', i.e. supplying electrical power for navigational equipment, lights, refrigerator, screen wiper, domestic water pumps, bilge pump etc. This prevents a situation where a long period between recharging makes it impossible to start your engine because the only battery is flat.

Electrical loadings vary considerably, depending on the type of boat and usage. As a rough guide, the Chloride Group suggest that where lighting and auxiliary loads are likely to discharge the battery in excess of 45 ampere-hours between recharges (e.g. 180 watts at 12 volts for three hours, or 360 watts at 24 volts for three hours), a separate battery should be used. This is even more important with a diesel engine, where the starting battery has to be able to produce a high current, sometimes as much as 1000 amperes, maintaining sufficient voltage to rotate the engine fast enough for it to start.

Diesel starting batteries have a different internal design to the

General Services battery, which is specified in respect of its amp-hour capacity. The special diesel battery usually has more plates per cell, of thinner section and minimal internal electrical resistance, with less acid capacity than the General Services type. It is specified to an SAE or IEC standard, which define the minimum voltage after a period of discharge at several hundred amps.

To withstand the pounding that a battery has to take in rough sea conditions, a heavy-duty marine type is built to higher standards than automotive diesel models, with stronger cases, better construction of plates, and better sealing of terminal posts etc. to prevent acid seepage. (Figure 1.33)

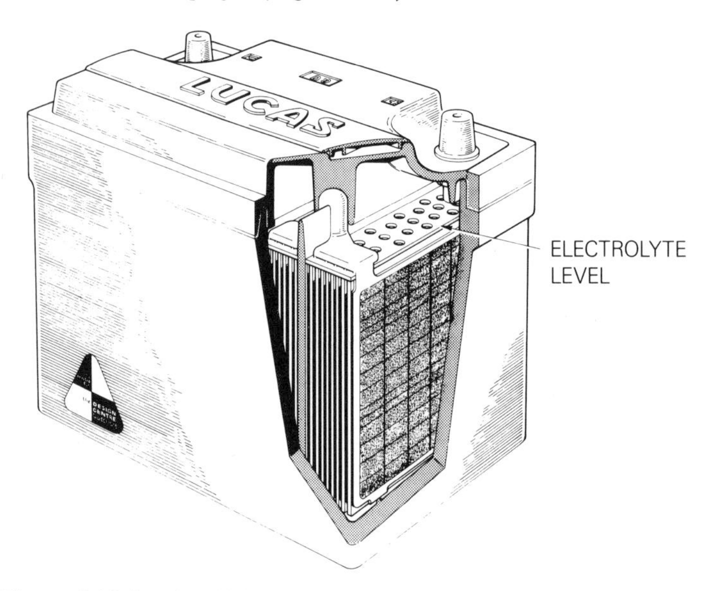

Figure 1.33 *Lead-acid battery*

Isolation of the two batteries, which will generally be recharged by the same engine-driven alternator, can be arranged in the wiring circuit, and an automatic method of preventing mutual discharge from both will be provided.

A system using blocking diodes or an electro-mechanical relay is specified according to the type of alternator. If this is *machine-sensed* the voltage output is controlled by a regulator set to a pre-determined voltage and sensed at the output terminals of the alternator. A relay is used as shown in the wiring diagram, Figure 1.34, and when the contacts are open the General Services loads will

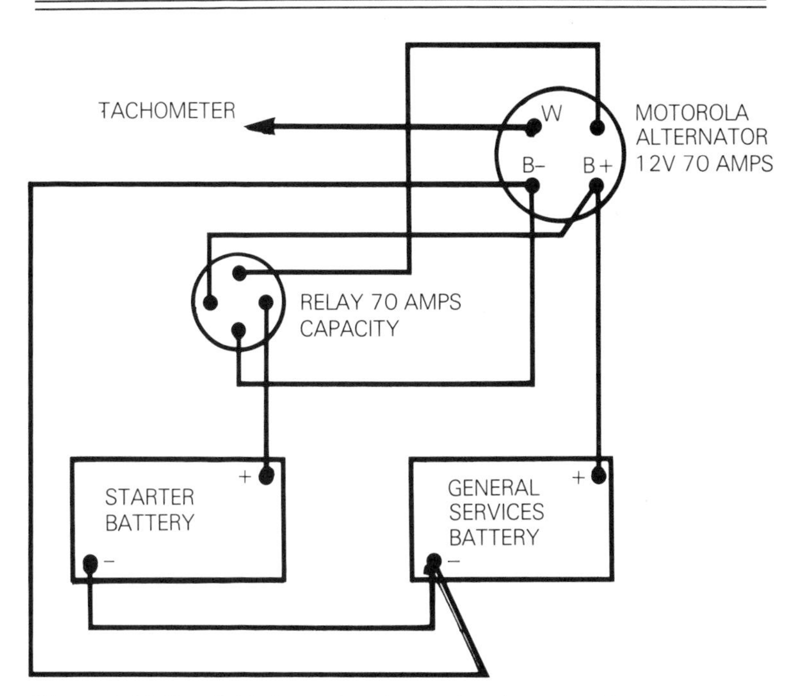

Figure 1.34 *Parallel battery charging system for a machine-sensed alternator*

not discharge the starter battery. When the engine is run so that the voltage at the alternator terminal reaches battery voltage, the relay operates, closing the contacts and paralleling the two batteries so that both are recharged.

A *battery-sensed* alternator has its regulator set to the same voltage as the 'normal' 14.2 volts for a nominal 12-volt circuit; but a separate sensing wire from the regulator is connected to a point close to the battery positive terminal which can be the output side of the isolating switch, as indicated in Figure 1.35.

Sealed Batteries

Several companies produce 'maintenance-free' batteries of which the cells cannot be topped up in service and the electrolyte, i.e. dilute sulphuric acid, is sealed in the case. Apart from the advantage of eliminating the need for topping up, the gassing which occurs when the battery is being recharged is reduced, preventing the release of explosive hydrogen gas in large quantities, and the possibility of an explosion when sparks are produced, as occurs

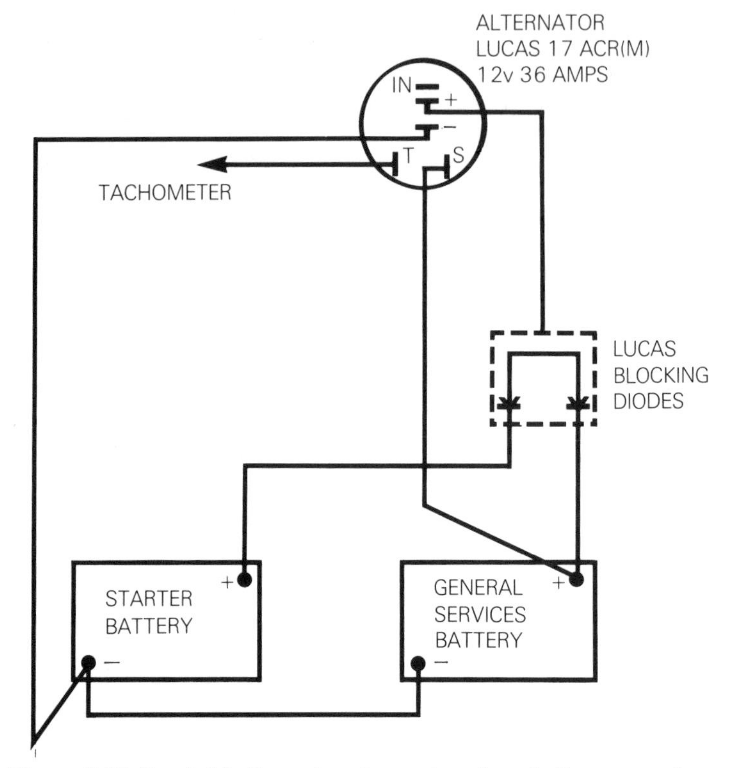

Figure 1.35 *Parallel battery charging system for a battery-sensed alternator*

when jumper leads are used across the terminals. A small vent is necessary to act as a breather to compensate for the effect of temperature changes. A device to separate the electrolyte liquid from gases is usually provided; this prevents the loss of electrolyte from the cells. A sealed battery minimizes corrosion by preventing the spill of acid. It also keeps the electrolyte in good condition by preventing the introduction of impurities when topping up, and it maintains the specific gravity of the electrolyte at the correct level.

Alkaline Batteries

These are more expensive than the commonly used lead-acid type and are larger and heavier. However, they will give longer life and are likely to be more reliable even when subjected to widely fluctuating loads and charging rates. Like the lead-acid type, the nickel cadmium battery releases explosive gas, and although alkaline the electrolyte is corrosive and will burn the skin.

Wiring Systems

The simplest wiring arrangement, often used for small launches and auxiliary engined sailing-boats, is *earth-return* – similar to that used in a car, where the negative current is conducted via the engine block to the casings of the starter, alternator and ignition system components of the petrol engine. The battery and any other electrical items, such as intruments and switches, have to be earthed to the engine. (Figure 1.36)

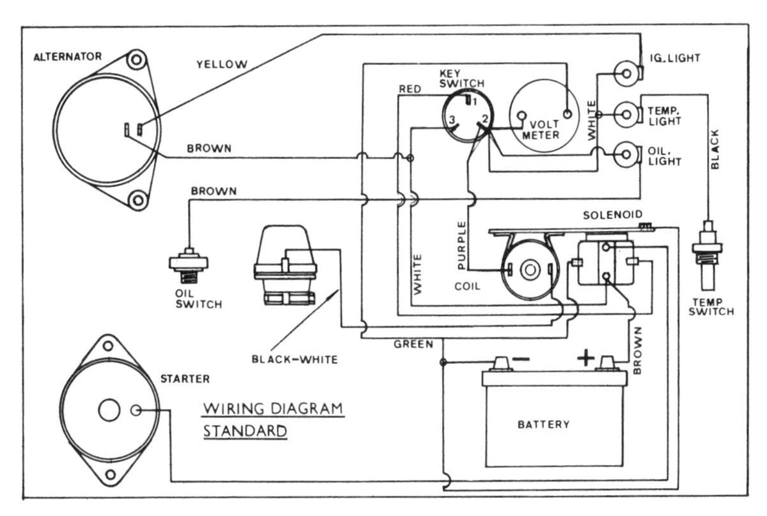

Figure 1.36 *Wiring diagram. Negative earth-return system Watermota petrol engine.*

The *insulated return* or *two-wire* system used with larger engines is superior because it does not rely on good external electrical contact to complete the circuit. Insulation failure is less likely to cause short circuits and leakage currents, which can cause electrolytic corrosion (see Chapter Ten) or radio interference. (Figure 1.37)

Radio Interference Suppression

Radio reception, also D/F equipment, can be virtually blotted out by interference from electrical sources on the engine, such as the

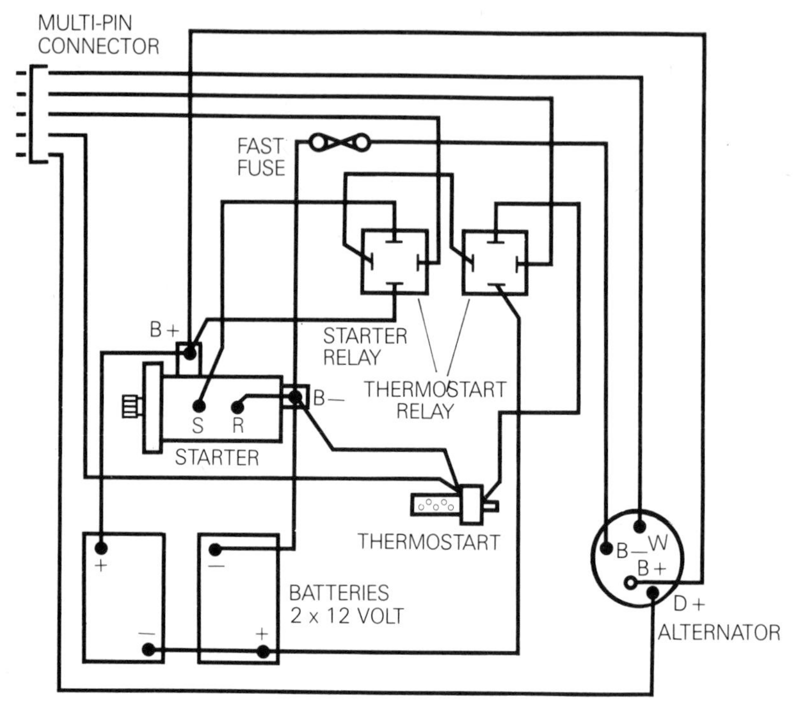

Figure 1.37 *Wiring diagram. Insulated return system for diesel engine.*

alternator, starter motor and regulator. Spark ignition engines have the additional problem of interference caused by the ignition system. This can be recognized by the loud crackling noise from the speaker and can often be dealt with satisfactorily by fitting a capacitor between the ignition feed side of the coil and earth, also a set of interference-suppressed plug leads with special plug caps and distributor end-caps. (Figure 1.38)

Electrical motors, thermostats, relays, switches and light fittings can also cause interference. Even the propeller shaft and metal items such as the rigging and mast or guard rails can be the source of interference. Preventing these problems is a specialized task and involves a special approach to the wiring system, entailing the use of metal conduit in some parts of the engine circuit, a specially adapted alternator, an earthing brush on the propeller shaft, and a common earth system to which the engine and other large metal objects are connected.

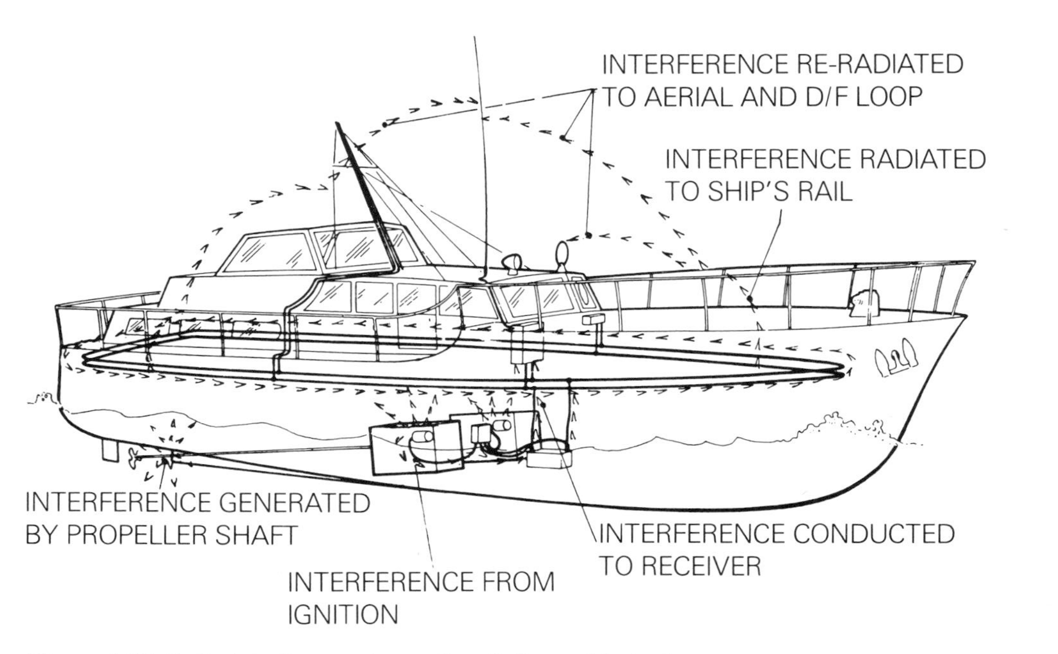

Figure 1.38 *Radio interference in small craft (Lucas Marine)*

Lucas Marine Limited recommend the following procedure to establish the source of interference:

1. With all engines, electrical gear and lights switched off, switch on the radio-telephone or direction-finding gear and note whether interference is present or not. If interference is present, the source is external to the craft and a change of moorings will effect a cure.
2. With the R/T or D/F set still switched on, switch on each electric motor in turn and note which motor or motors causes interference.
3. Switch off all electric motors and start engines. Run engines at sufficient speed to ensure that the generators are charging the batteries, and again note if interference is present.
4. Stop engines and disconnect main output and field leads from engine-driven generators, after carefully noting cables and terminals to facilitate correct reconnection. Start engines, if they are spark ignition petrol or paraffin engines, open and close throttles and note if interference is present. If the ignition is causing interference this will be apparent with the rise and fall of the frequency in synchronization with the opening and closing of the throttle. Slow engines to idling speed and engage the propeller drive. Again accelerate and decelerate the engines and note whether interference is produced by the propellers and propeller-shafts.
5. Disengage propeller drive and stop engines. Reconnect main output and field leads of generators to correct terminals as previously noted. Switch off R/T or D/F set.

A comprehensive kit of parts to suppress their alternators is provided by Lucas Marine and is illustrated in Figure 1.39. The main feature is the suppression box which houses the voltage regulator, surge protection and radio interference suppression filter network. This equipment will allow a very high standard of suppression. For pleasure-craft a slightly lower standard is often acceptable and this is obtainable by fitting chokes and condensers to the alternator, as shown in Figure 1.40. This arrangement can be accommodated on the alternator, and in the illustration a Motorola unit is shown with the necessary components fitted. A kit of suppression components can be obtained from the alternator supplier or the engine manufacturer's distributors.

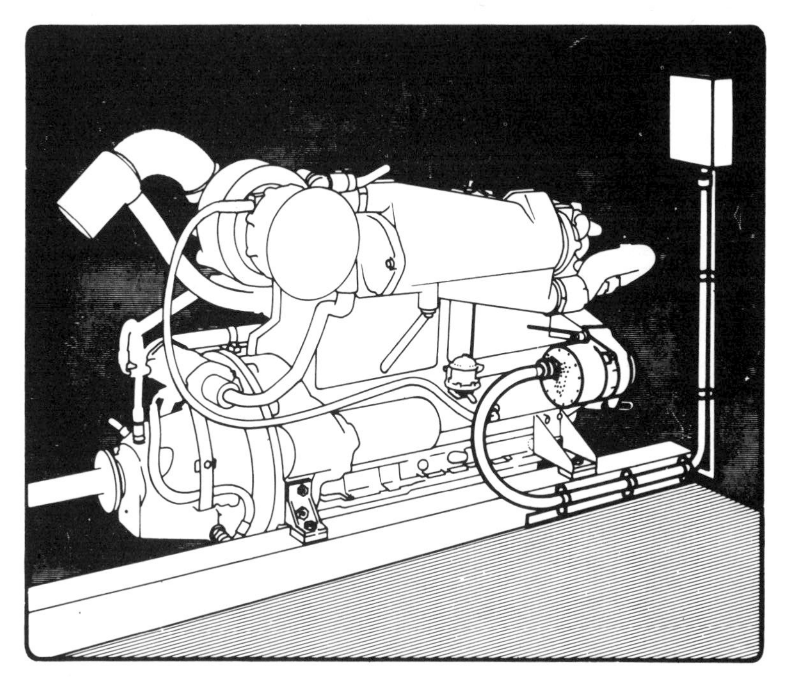

Figure 1.39 *Radio interference suppressed charging equipment (Perkins)*

Figure 1.40 *Interference-suppressed alternator*

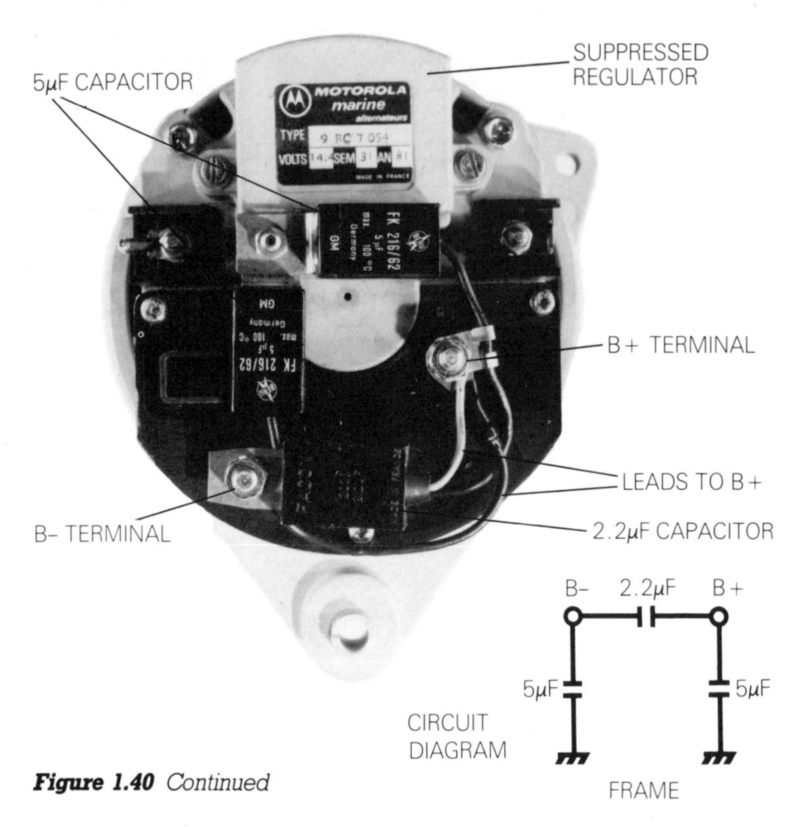

Figure 1.40 *Continued*

Instrumentation

Opinions vary as to what instrumentation is really necessary to allow the condition of a petrol or diesel engine to be monitored. Obviously the needs of a small auxiliary engine are very simple compared with those of a highly tuned racing engine. It is generally agreed, however, that an oil-pressure gauge is the most necessary item for any engine. An oil warning light can also be used to show when the lubricating-oil pressure is below the lowest permissible level to maintain satisfactory lubrication.

An ammeter or a charging light was once considered essential, although nowadays a voltmeter to indicate the state of the battery is generally preferred.

Revolution counters have always featured in marine instrument panels for the larger engines and are useful when setting the throttle to a predetermined level – for example, the cruising setting or the engine speed to obtain the most nautical miles per gallon or minimum gallons per mile, according to the consumption of the engines.

Water temperature gauges measure the fresh-water temperature in the cylinder-head of a heat-exchanger or keel-cooled engine. Overheating caused by an engine malfunction, a stuck thermostat or loss of coolant will show up on the temperature gauge. You may not know what the fault is, but the gauge will show that the engine should be stopped as soon as possible. Of course, if you are in a potentially dangerous situation you may decide to reduce the engine speed as far as possible until you can heave to, and lift the engine cover.

Some engine-builders recommend lubricating-oil temperature gauges in addition to an oil-pressure gauge, because of the shortcomings of the latter, which cannot really tell you if all is well throughout the engine speed range. If the oil is seriously overheated a bearing failure could result, and although the pressure gauge or warning light might indicate deficient pressure at idle speed, at higher engine speeds it could still register normal pressure. Figure 1.41 shows a typical panel.

Figure 1.41 *Instrument panel. Incorporates VDO Electrical Instruments and audible/visual warning device (Perkins).*

1 Engine oil pressure gauge
2 Audible alarm
3 Tachometer
4 Water temperature gauge
5 Voltmeter
6 Oil pressure warning light
7 Water temperature warning light
8 Oil temperature warning light
9 Alternator warning light
10 Engine heat/start switch
11 Panel light on/off switch

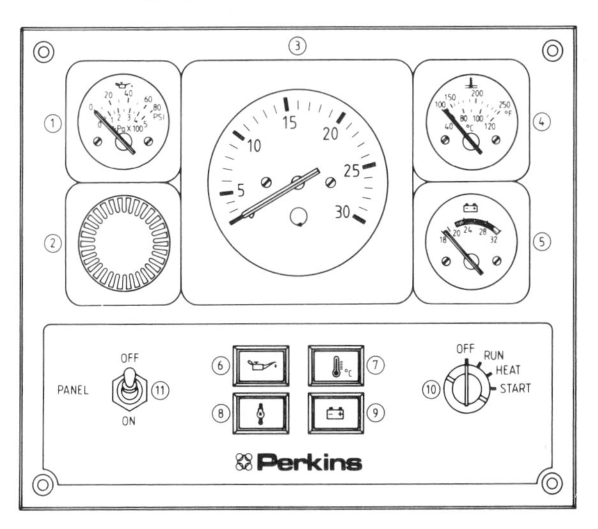

Turbo-charged engine instrument panels are sometimes equipped with boost gauges which register the air pressure after the compressor. Owners sometimes worry about slight variations of boost pressure in a pair of engines; this may or may not be significant, but in many cases slight variations in the accuracy of the gauges or in the pipework can be the cause.

Figure 1.42 *Twin station instruments (Perkins)*

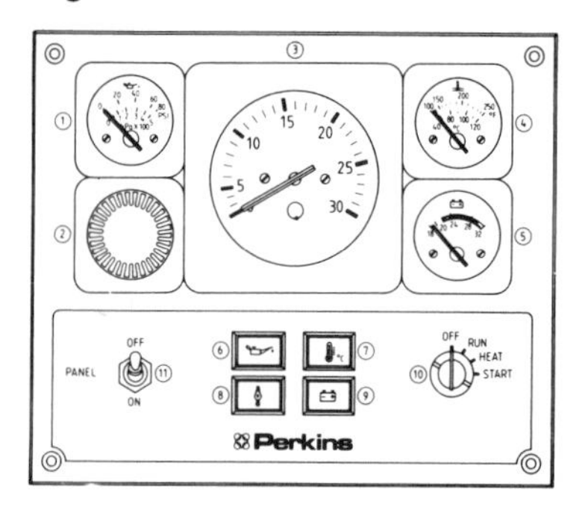

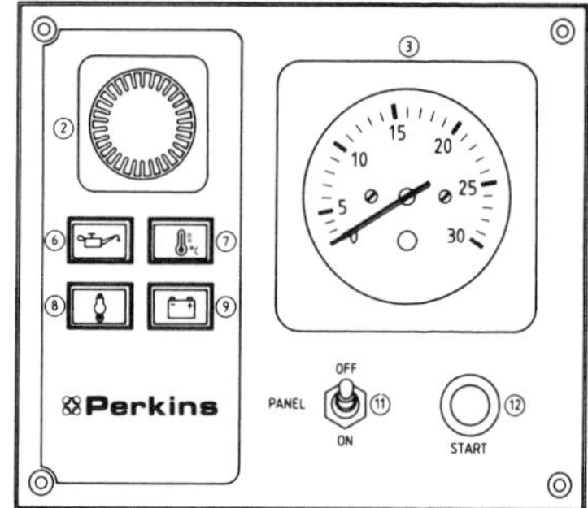

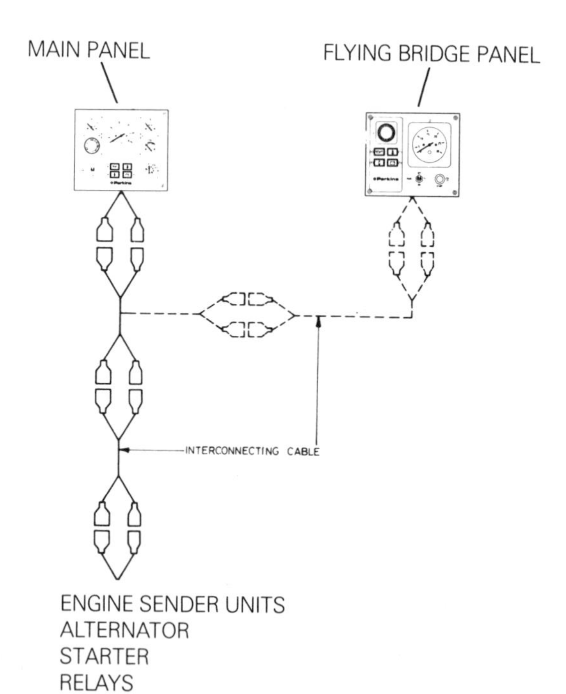

V8 engines with two banks of cylinders sometimes have instruments which monitor both, so that a panel for a twin-engine installation resembles the cockpit of an airliner, especially if matching navigation instruments are included in the grouping.

The simplest instruments are mechanical. The tachometer is driven by a cable similar to that of a car speedometer; pressures are measured directly by gauges, increasing pressure straightening a bent tube linked to the pointer; temperatures are measured by capillary tubes. Electrical instruments are more sophisticated and have several advantages over the mechanical type. With these there is no limitation on the distance of the instrument panel from the engine, which has 'sender' units and switches screwed into the monitoring points. The engine speed is measured by variations in the voltage from the alternator at the special terminal provided. Electrical instruments usually have plug-in multi-point connectors to join the wiring harness on the engine to the loom which runs to the instrument panel. Electrical instruments allow a second, or even a third, panel to be provided, with a single set of sender units and switches on the engine. (Figure 1.42)

2 Preventive maintenance

Preventive maintenance can be regarded as an insurance against trouble and a means of boosting one's confidence before taking the boat out to sea. Very few engine breakdowns are entirely beyond our control, and the majority of engine stoppages could have been avoided by a good maintenance regime – which essentially consists of keeping a weather eye on the condition of the engine and its services while carefully carrying out the prescribed maintenance tasks.

Car-owners may concentrate on maintaining the brakes, tyres, lights and other safety features but neglect the condition of the bottom radiator hose, which is then liable to burst on a Sunday evening in the middle of Dartmoor. We can't afford to risk this sort of thing happening to a boat engine, as the consequences may obviously be more serious. However, the modern diesel or petrol engine and outboard motor are pretty reliable machines if looked after properly, and some of the basic principles apply to all three forms of power.

Cleanliness is said to be next to godliness, and this is certainly the most important principle to be observed in operating engines of all types. The golden rule is to make sure the engine has supplies of clean air, fuel and lubricating oil. We do this by following a strict regime as described in the following pages.

Petrol Inboards

Most petrol engines used in pleasure-craft are based on car engines, so that the maintenance principles of the basic engine are very similar to those described in the car-owner's handbook. The cooling system is one important difference, however, and so is the transmission.

The operating conditions are also very different, with the crankshaft tilted back at 15 to 20 degrees, heeled over during tight turns in a ski-boat, splashed with corrosive bilge water, and left for days or weeks while condensation forms in the exhaust or induction manifolds and runs into the cylinders. It is a wonder that the engine runs at all; indeed sometimes it won't even start because the battery doesn't like the marine environment.

Servicing and preventive maintenance operations described below are intended to keep the engine in good order in spite of the hostile environment. The frequency with which they are carried out is dealt with later in the chapter.

Fuel System

Keeping dirt out of the fuel system is important. Blocked jets will necessitate cleaning out the carburettor, and this is a job to be avoided except when overhauling the fuel system at the end of the season. Regular attention should be paid to the filter screens at the fuel lift pump and carburettor intake. (Figure 2.1) Clean with a soft brush dipped in petrol. A filter located in the filler tube will help to avoid a build-up of dirt in the tanks.

Water is another substance to be avoided. It may be caused by condensation in the tank, but however it got there you should have a water-trap or other means of separating it out, because a little in the carburettor can stop the engine. Regular attention to the water-trap – draining off water which settles in the bottom – is necessary. (Figure 2.2)

Yet another hazard is the gummy residue which forms in the fuel lines and carburettor during long periods of inaction unless the system has been drained and flushed out with clean fuel.

Ignition System

Chapter Four deals with tuning the ignition system. This entails removing the spark-plugs for cleaning and resetting, checking and possibly renewing the points in the distributor, and checking the condition of the plug leads and other wiring as well as the ignition timing. These are also preventive maintenance jobs since they have the effect of putting the ignition system in good order for about 250 hours of service.

Before the next tune-up is due, however, the plugs may benefit

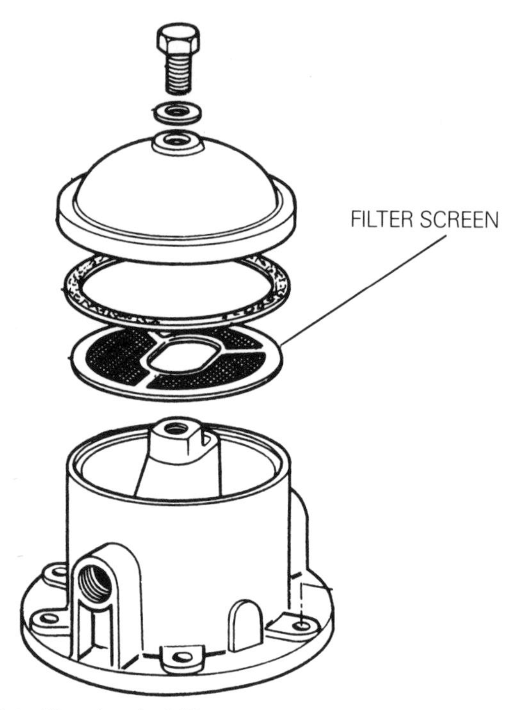

Figure 2.1 *Cleaning fuel lift pump*

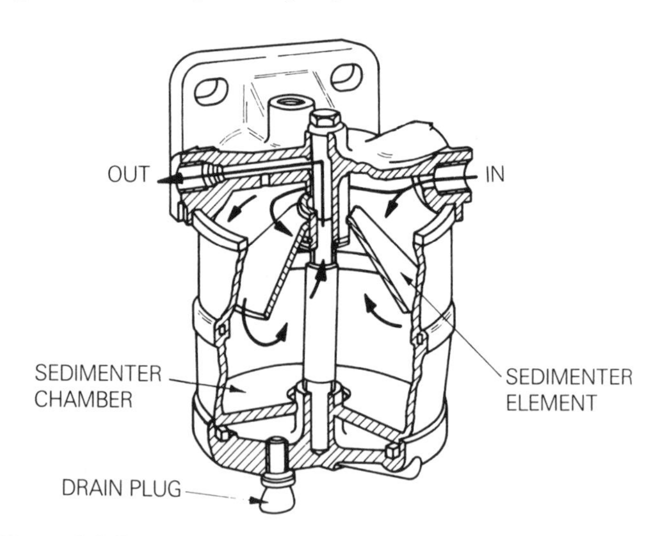

Figure 2.2 *Fuel/water separator. CAV 'Sedimentor' for petrol or diesel engines.*

from cleaning and resetting the gap. Remove the plugs one at a time to avoid getting the leads crossed when refitting. Blow out the carbon, clean the points with a small wire brush, and check the gap, which is usually between 0.025 and 0.028 in. (0.60 to 0.70 mm.). Bend the side electrode, not the centre one, if you have to adjust the gap. Refit carefully, making sure the washer is in position, and clean out any dirt in the recess without letting it drop into the cylinder. Do not use excessive pressure on the spanner – use a new washer if the old one has flattened and will not seal properly.

Before cleaning the plugs, take a look at their firing end as this will show whether your tune-up was satisfactory and the fuel mixture correct. Normally there should be powdery deposits, coloured brown to grey, and the electrodes should be slightly eroded by the millions of sparks generated. A too-rich mixture in the carburettor will cause dry black deposits on the plugs. If the plug end is coated with an oily deposit, and especially if you have had to top up the lubricating-oil frequently, you probably have worn piston rings, so that the engine will soon be in need of an overhaul. The sparking-plugs are a good indicator of the engine's condition.

Keeping the leads, coil, distributor cap and plugs free from dampness will assist starting and avoid misfiring. Dry off the vulnerable parts of the ignition system with a clean cloth before starting up on a cool, damp morning, and spray occasionally with a moisture inhibitor such as 'Dampstart' or W.D.40.

Electrics

Starter Motor

The starter motor must be kept in good order. Any tendency for the pinion to stick in mesh with the gear ring on the flywheel, or poor performance when the battery seems to be well charged up, must be investigated before it lets you down. Don't strain the starter by keeping the engine turning over more than about five seconds – unless it is a dire emergency, of course. If you put your hand on it and feel how quickly it heats up you will soon learn to exercise restraint.

With Bendix drive models (as described in Chapter One), the pinion may stick to the shaft if rusting occurs. A light penetrating oil applied at the end of the season will help to prevent this.

This will generally entail removing the starter by undoing the flange screws and withdrawing it from the flywheel housing. When you do so have a look at the gear ring teeth on the flywheel. If you have had problems with pinion engagement and the gear teeth are badly worn, the only cure is removal of the flywheel and a new ring shrunk on – definitely a job for the engine-dealer even if you decide to remove it from the engine yourself. To do this you jack up the aft end of the engine, detach the propeller-shaft coupling from the transmission flange, remove the rear engine holding-down bolts and then, having disconnected pipes, controls and wiring connected to the transmission, remove this by undoing the retaining screws from the attachment flange so that you can pull it away from the engine flange.

The transmission can then be removed, revealing the drive coupling and flywheel. A heavy transmission will require shear-legs or at least several willing helpers to lift it clear of the engine. The retaining screws in the centre of the flywheel are removed and it can then be detached from the crankshaft 'palm', i.e. flange. Some flywheels have tapped holes in the centre so that you can screw in the attachment screws and thereby push the flywheel off the crankshaft palm spigot. Failing this, screwdrivers wedged around the periphery and gentle taps in the centre with a hide hammer are necessary – and if this doesn't work you may think you should have called in the engine-dealer after all!

Back to the starter: the main thing to watch is that the electrical connections are sound – not only the cable supplying current from the battery, but also the earth return through the flange of the machine to the engine casing, and the earth connection to the engine. If the starter is wired on the insulated return system there is, of course, no need to have good earth connections but all terminals must be bright, tight and free from corrosion.

After several seasons' use the efficiency and reliability of the starter may be improved by a bench overhaul, as described in Chapter Five.

Alternator

Little routine maintenance is required. Electrical trouble-shooting is described in Chapter Three, but mechanical problems are generally confined to worn bearings, easily detected by noisy operation – which can also be caused by a worn driving-belt, loose pulley or bent fan. While you can deal with the belt, pulley or fan, a bearing change is best done in a workshop with the necessary press and spacers.

Battery

Make sure you have a reliable battery, one that won't let you down just when you need to start the engine.

It is better to have two separate batteries – one for the engine starter and one for the navigation lights and other auxiliaries. If you don't have two it is something you can arrange (providing you can find the room) during the winter lay-up. A change-over switch or wiring circuit enabling a single alternator to charge both batteries is necessary for a single-engined boat. With a twin-engine installation you can charge the starter battery from one engine alternator and the auxiliary battery from the other. Whichever way you do it, make sure that you do not inadvertently disconnect the alternator from the battery while the engine is running. This is sudden death for most alternators.

Routine battery maintenance consists mainly of checking the electrolyte level in the cells and topping up when required with distilled water. Use clean tap water or rain water if you do not have distilled water available. If you need to top up frequently you may be overcharging due to incorrect setting of the regulator voltage, or maybe the battery gets too hot because it is too close to the engine. Keep the terminals free of corrosion by coating them with Vaseline or grease. The battery should be in a separate compartment, ventilated to allow the hydrogen gas (emitted by the cells during charging) to escape to the atmosphere.

When you need to check the condition of the battery you can either measure the voltage of each cell or check the specific gravity of the electrolyte, i.e. the dilute hydrochloric acid with which it is filled. The voltage of each cell should be just over 2 volts so that the total battery voltage is approximately 14 volts for a nominal 12-volt battery when fully charged. The hydrometer draws a sample of the electrolyte from the cell and indicates the level at which a float stands in the tube. The float is calibrated to indicate the specific gravity reading and often scaled to show the condition, i.e. fully charged, partly charged or discharged. (Figure 2.3)

To prevent your battery from becoming discharged when the boat is out of use, the positive lead should be disconnected. If you have no main switch the terminal can be prised off the positive terminal post after loosening the clamp. This is a chore – it is worth while fitting a battery switch, making sure it is accessible.

When the battery needs charging, a trickle charger connected to

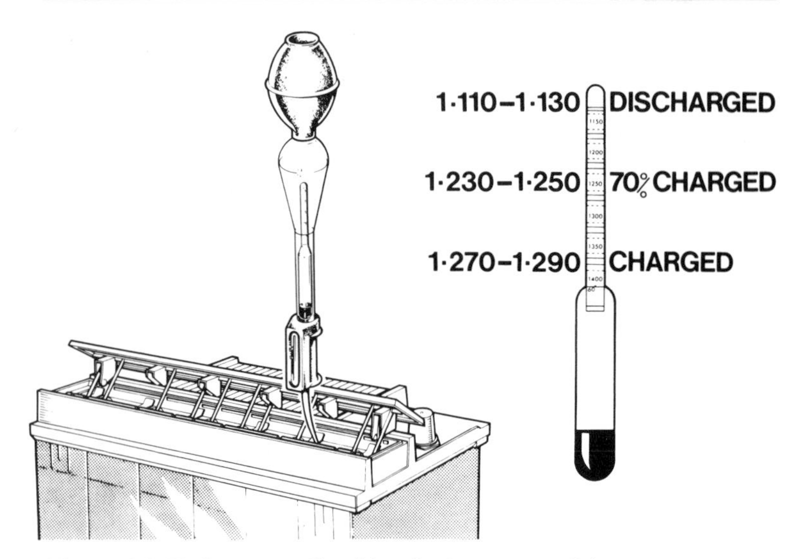

Figure 2.3 *Hydrometer. Checking the battery condition.*

the mains supply is the easiest method, but do not be tempted to leave this connected for several days at a time – overcharging can damage the battery. If the battery is flat and you don't have mains supply for a trickle charger, you can use jump leads to connect a spare battery in order to get the engine started. The alternator will regulate its charge to the battery condition, and a few hours' use should restore the battery to a fully-charged condition. If it doesn't there is something wrong with the battery, its connections, the alternator, regulator or wiring. Checking the fuses, battery cable connections, earthing and condition of the charging circuit wiring is straightforward, and an occasional look at these items should figure on your routine maintenance check list. If these checks don't bear fruit the alternator and regulator should be checked – preferably *in situ* – by an auto electrician.

Lubrication

Change the lubricating-oil at the prescribed periods. Always drain the oil when the engine is hot so that it flows through the oil-ways. If you turn the engine over by hand or on the starter with the ignition switched off, you can squeeze the last few drops out of the system after draining the sump and removing the old oil filter. As there is usually no room to get a receptacle under the sump to receive the old oil, it is necessary to pump it out with a hand pump. This is either built onto the engine or used with a length of rubber or plastic tube for attachment to a drain connection. On some engines the dipstick tube extends to the bottom of the sump and the rubber pipe is attached to the top of this tube, to allow the oil to be pumped out.

Make sure you drain all the oil out of the sump, as a pint or so left in it will retain impurities and you will suck them back into the engine again when you start up. (Figure 2.4)

Figure 2.4 *Draining the oil sump. If the engine is not fitted with a drain pump access is usually provided via the dipstick tube or a special drain connection.*

To renew an oil filter canister unscrew it from the filter head or cylinder-block. If it is too tight, use a chain wrench as illustrated in Figure 2.5. This is purpose-made for the job and is available from most tool-shops and motor-stores. Discard the old canister, clean the joint surface and fit a new rubber joint ring, making sure you remove the old one. Lubricate the joint surface before screwing the

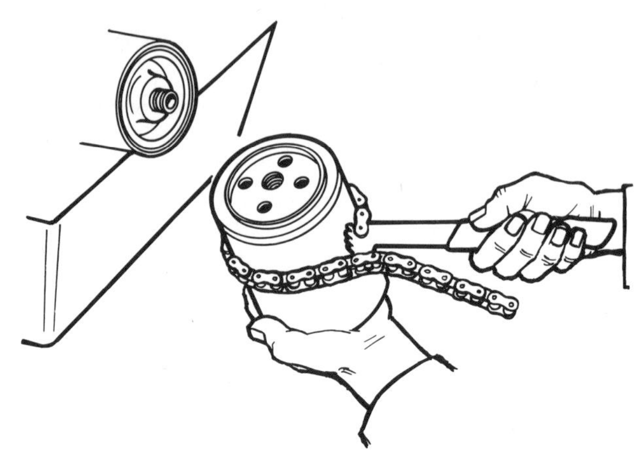

Figure 2.5 *Removing oil filter canister*

replacement canister into position. Do not overtighten; a torque of about 15 lbf. ft. (2 Kgf. m.) is adequate. Less scientifically, allow about half a turn after you feel the gasket resistance.

If your engine has an upside-down oil filter position, make sure that your replacement canister is to the correct specification: these sometimes have a stack pipe inside to prevent the oil draining out when the engine stops, so delaying the supply of oil to the engine bearings when the engine is started.

After starting the engine with the new filter, examine for leaks before running at full speed.

Check the sump oil level and top up as necessary. Incidentally, keep the oil level between the 'full' and 'low' marks on the dipstick. Overfilling to be on the safe side does not pay because the extra oil in the sump will soon be consumed – it will be churned up by the con-rods and an excessive amount will be thrown onto the cylinder walls and burnt.

Air-cleaner and Breather

Maintenance is straightforward. With the paper element type you throw away the old one and fit a new part. Never try to clean and re-use the old element. If you have the simple metal screen type of air-cleaner wash this in petrol or cleaning solvent using a clean paint-brush.

Make sure the rubber breather pipes are not cracked or

perished, and wash out any part of the system where oil has collected.

Check that the air-cleaner fits squarely to the carburettor flange and renew the joint if necessary.

Cooling System

It is possible that after a long period of service the heat-exchanger tube stack will become fouled so that the engine overheats. On some exchangers with detachable tube stack this can be extracted without removing the heat-exchanger from the engine. (Figure 2.6)

Removing the end-covers allows access to the tube stack, which can then be eased out of the body. Corrosion can occur on the sea-water side, which is generally the inside of the tubes – fresh water flowing over the outside. Pushing a length of steel rod through the tubes will generally clear any blockages. You should push the rod in the *opposite* direction to the sea-water flow – you can work this out by checking which end-cover is connected to the sea-water

Figure 2.6 *Removing the heat-exchanger tube stack*

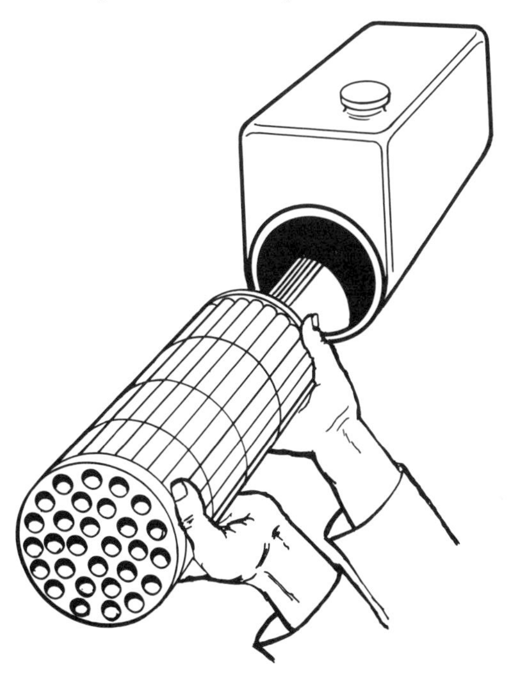

pump delivery. Don't use too much force with the ram-rod – the tube material is easily damaged. A first-class job can be made by boiling the tube stack assembly in caustic soda solution, providing the baffles are not made of aluminium. Before refitting, clean the end-covers and renew any rubber washers or joints that have hardened, perished or are damaged.

Lub-oil and transmission coolers where sea water passes through the tubes can be cleaned in the same way as described for the heat-exchanger. If you have the soldered-up type you can use the boiling caustic soda method unless the manufacturer objects to such drastic treatment – when in doubt ask your engine-dealer.

The fresh-water part of the cooling system includes the pump – generally the same type as is fitted to a car engine belt driven from the crankshaft pulley, which usually drives the alternator – occasionally a dynamo – at the same time. The cylinder-head and cylinder-block water-jackets, and possibly the exhaust manifold jacket, complete the circuit. Very little maintenance is required on these items because the pump usually has sealed bearings so that you cannot fill with grease. If the pump gets noisy (remove the belt to check this properly) or if you can wobble the pulley, then the bearings are about to fail; the best remedy is a service exchange water pump. These pumps generally last longer on a boat engine because they do not have to carry a fan and, unless yours is a racing engine, the maximum running speed will be lower than for a road vehicle engine.

Particularly if you experienced a fresh-water leak and have had to top up with sea or even brackish water, drain and flush the system. This is particularly important if you have any aluminium components in contact with the cooling water, because corrosion will soon eat through the thin casting walls.

Sea-water Pump

The sea-water pump should need no routine attention except a visual check for leakage from the seal at the same time as you give the other external parts of the cooling system a glance, to see that all is well.

If the seal is leaking – and this may happen after a prolonged period of service, especially when running close to the shore in sandy conditions – it must be replaced by a new assembly.

The counterface on which the seal runs should also be replaced if the seal is of the face-sealing type. A spare seal should be included in the on-board spares kit and, of course, replaced when fitted to the pump. See page 111 for further information.

The pump bearings are usually grease-packed and require no maintenance. If the grease seal fails and sea water enters, the bearing will not last many hours and the bearings must then be pressed out and replaced. This is a job for the workshop, so the boat will be immobilized until the job is done unless you fit a replacement pump. If you have conventional bearings requiring grease lubrication, obviously this has to be regularly attended to with a few strokes of the greaser or turns of the grease cap. A good-quality waterproof pump grease is required.

The pump impeller requires no maintenance but the blades can take a permanent set, or stick to the pump casing, if the engine is not run for several months. It is best to remove the impeller, but if this is not possible you must check the impeller before starting the engine after a lay-up. If the blades have taken a set, immerse the impeller in very hot water for a few minutes – this will straighten things out. Give the impeller blades, pump housing and shaft splines a light coating of grease before replacing. (Figure 2.7)

Figure 2.7 *Sea-water pump. Remove end plate to check condition of impeller and lubricate rubber blades.*

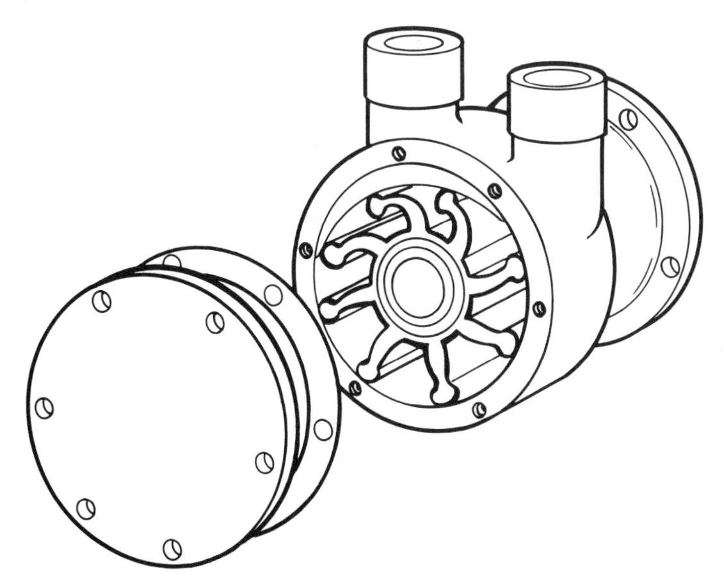

If you have had an impeller failure you may be missing one or more of the blades or at least parts of them. These may disappear into the pipes and reach the heat-exchanger end cover, but usually they cannot get any further; they must be found and removed.

One thing that shortens the life of impellers is running with excessive suction such as is caused by blockage of the sea-water inlet. It is necessary, both in order to protect the water pump and to prevent engine overheating, that the inlet strainer should be regularly checked.

One final point: rubber impeller pumps are self priming – there is no need therefore to remove the screw holding the cam, which has been mistaken for a priming screw. If you do take it out for any reason, coat it with jointing compound before replacing it, to avoid a water leak past the threads.

The Maintenance Routine

It is a good idea to write your maintenance procedure onto a card and keep this in the wheel-house or engine compartment. Periods between oil changes and other functions vary according to the engine-manufacturer's recommendations, but a general schedule might be prescribed as follows:

Petrol Engine Maintenance Schedule

Daily (or between each 'leg' of a passage)

With engine stopped; check and replenish as necessary:
Lubricating-oil – dipstick level; Fresh Water – level in header tank; Transmission oil – dipstick or level plug – Fuel level – gauge or tank dipstick; Stern-gland lubrication – oil level in container.

With engine idling on mooring:
Give the stern-gear greaser one turn.
Check engine, water, exhaust, fuel and oil pipes, hoses and connections for leakage.
Make sure sea water is discharged overboard with engine idling.

Monthly

In addition to daily check:
Check alternator drive belt tension.
Readjust alternator position on its bracket as necessary.

Clean water-trap in fuel pre-filter.
Check exhaust pipe connections to transom fitting and silencer.

Every 150 hours (or as prescribed by engine supplier)

Drain and renew engine lubricating-oil.
Renew lubricating-oil filter canister.
Clean fuel pre-filter screen and drain water-trap.
Clean air-intake gauze.

Every 400 hours or 12 months (whichever is sooner)

If at end of season and boat is to be laid up for the winter, proceed as described for 'Winter Lay-Up' in Chapter Ten.

If the boat is to remain in service:
Renew air-cleaner paper element (if fitted).
Tune engine (see Chapter Four). Replace plugs, points etc. as necessary.
Change gearbox oil.
Check and adjust valve clearances.
Make close examination of all external hoses, pipe connections, clips and cooling system components for signs of corrosion, deterioration or leakage.
Check condition of belt, and replace if necessary.

Diesel Maintenance

The procedure is the same as for the petrol engine except that you deal with the fuel injection system instead of attending to the ignition. Items specific to the diesel engine are as follows:

Every 400 hours or 12 months (whichever is sooner)

Renew the fuel filter element.
Service the atomizers.

Checking Valve Clearance

Some engines have the valve clearance specified with the engine cold and others with it hot. Expansion of the tappet and valve when the engine is hot reduces the clearance, so make sure you carry out this check at the specified condition. The actual clearance is often given on a plate attached to the rocker box but if you are unable to find the specified figure, ask the engine dealer – and if there isn't one available, 0.30 mm. (0.012 in.) with a cold engine is a common setting.

Figure 2.8 *Adjusting tappet clearance. Overhead-valve engine.*

Proceed as follows after removing the rocker box for an overhead valve engine.

Using a ring-spanner and a screwdriver to adjust the tappet screw and locking nut, slide a feeler-gauge blade (of the appropriate thickness equal to the specified clearance) between the valve rocker arm and the end of the valve stem. Adjustment of the screw should be made until the feeler-gauge slides easily between the valve stem and rocker arm. Tighten the locknut carefully so as not to affect the adjustment. (Figure 2.8)

Replace the rocker box joint if this is not serviceable.

3 Trouble-shooting

If you do your preventive maintenance well, the engine is obviously less likely to let you down than if you neglect it. However, some mishaps are almost impossible to prevent, so we must learn to deal with the most likely ones – and some of the less common problems as well.

Sooner or later you are likely to run out of fuel or start up with the fuel-cock turned off so that air gets into the diesel fuel lines; it is no problem to bleed the air out of the system if you know the drill.

Petrol engines can get dirt in the carburettor, or other ignition system problems. Again, this is no problem if you know what to do and have rehearsed the procedure – particularly with the engine hot.

Built-in Problems

Some boats have 'built-in' problems caused by engine installation faults. These are generally one-off 'specials'. It is less common for production boats to have these problems as, even if some were present in the first boats built, owners' complaints would soon lead to modifications in subsequent craft.

Trouble-shooting installation faults consists of diagnosing the cause of the trouble and then correcting it, which in many cases means working to the engine-builder's or equipment manufacturer's recommendations.

Some of the more common problems which can occur because of faulty installation or unsuitable equipment are as follows:

Symptoms	*Cause*	*Cure*
Engine labours and does not produce full power or maximum revolutions. (Diesel or Petrol)	Inadequate supply of air to the engine intake caused by inadequate ventilation.	Fit specified ducting or air vents. (Figure 3.1)
	Note: Check if this is the fault by raising the engine compartment hatch cover while engine is running; if revs increase this is the cause. If not, check:	
	1. Size of propeller(s) (diameter pitch, number of blades, type of propeller). Refer to boat-builder or propeller specialist.	Replace by correctly specified prop(s).
	2. Back-pressure in exhaust pipe. If well in excess of engine-maker's standard, this may be the cause.	Increase diameter of exhaust pipe or larger size of silencer.
	3. Incorrect reduction gear ratio.	Fit correct reduction gearbox.
Vibration from propeller-shaft (Diesel or Petrol)	Propeller damaged or shaft bent. Misalignment of engine to propeller-shaft or absence of flexible coupling.	Rectify faulty item. Check alignment after boat has been in water several days. (Fig. 3.2) Fit flexible coupling if stern-gear layout requires this. (See Chapter Ten)
Engine Vibration (Diesel)	Absence of flexible engine mounts or unsuitable type.	Fit correct mountings. (See Chapter Ten)
Excessive engine noise.	Inadequate soundproofing material in engine compartment. Gaps transmitting noise from compartment.	Apply soundproofing material, seal up gaps. (See Chapter Ten)
Propeller noise.	1. Propeller out of balance.	Balance propeller.
	2. Propeller blades too close to hull.	Fit smaller prop if possible without imparing performance or manoeuvrability. Try different type of prop – four

Symptoms	*Cause*	*Cure*
		blades instead of three, for example. Otherwise alter shaft line – a difficult operation. (See Chapter Nine)
	3. Poorly supported propeller-shaft, inadequate outboard bearings.	Fit necessary bearings and shaft brackets.
Fuel starvation or air in fuel when tanks only partly full.	Poor pick-up pipes in tank or inadequate connecting pipes with multi-tanks. Disconnect supply pipe and run engine from temporary tank to check if the system is faulty.	Rework the plumbing as necessary.
Water in Fuel.	1. Deck filler not watertight. Note: some fillers leak only when side decks awash.	Fit sealed filler.
	2. Breather pipe lets in water.	Modify pipe end.
	3. Inadequate (or missing) water-trap.	Use engine-builder's recommended water-trap.
Hydraulic lock in cylinders.	Sea water enters cylinders from exhaust pipe.	
	1. Inadequate pipe fall from engine to transom.	Improve pipe run or fit waterlock system.
	2. Fault waterlock system. (See Figure 10.8)	
Rain or sea water in engine compartment via ventilating system.	Poor design or siting of intake vents.	Modify intakes.

Other shortcomings in the machinery installation may not lead to problems in the first few weeks of use but can cause trouble later. Safety standards as well as mechanical or electrical reliability will be affected by the following:

Poor service access to engine(s).

Untidy, exposed fuel lines, electrical wiring, control cables etc.

Figure 3.1 *Air inlet ducting (Perkins)*

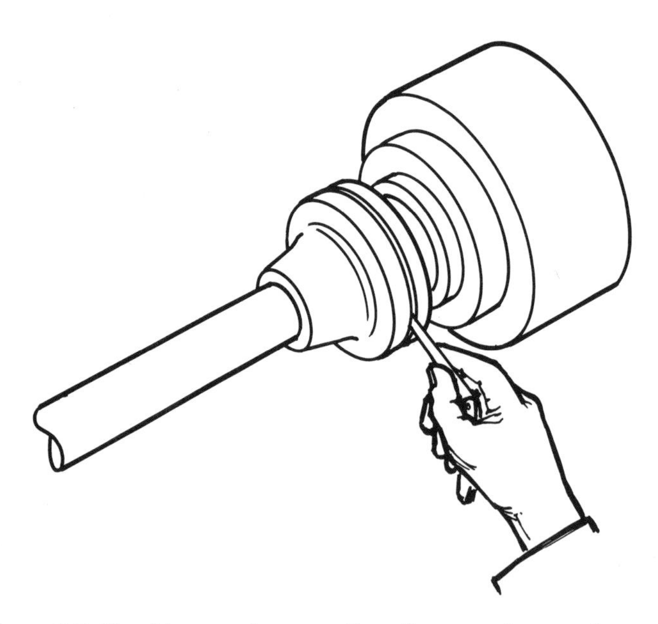

Figure 3.2 *Checking gearbox coupling alignment. Support the propeller shaft close to the coupling to allow for bending, also deflection of the stern tube bearings.*

Insecure fuel-tanks; unsuitable tank material liable to corrode; no shut-off cocks; no contents gauge; no baffles; plastic piping to engine.
Single battery for engine starting and lighting, etc; inadequate capacity; sited in engine compartment; not properly secured; inadequate cable sizes; no master switch.
No sea-water strainer; inaccessible sea-cocks.
Inadequate bilge pumps; no fire-extinguisher; no flame trap on carburettor.

'On-board' Spares

In order to be able to deal with problems as they arise, you need to carry some spare parts for the engine and also miscellaneous items for minor repairs to piping and those fittings that only seem to need attention when you are away from the mooring.

Owners' ideas on what to carry on board vary from a trunk full of tools and engine spares, weighing at least a hundredweight, to a pocket set of spanners and a roll of insulating tape.

Engine manufacturers will supply a kit of parts to cover most engine problems that you can deal with at sea. Some additional items allowing you to repair the installation pipework are advisable, however. If you decide to make very long voyages, or spend the summer cruising an area where you won't find a dealer with the right spare parts, a more comprehensive inventory is needed.

The following items are suggested for both diesel and petrol engines:

Set of engine joints and gaskets
Set of rubber hose connections with stainless steel clips
Sea-water pump repair kit (impeller and seal assembly)
Sacrificial anodes (if used on your engine)
Alternator and pump drive belts
Alternator
Fuses
Voltage regulator (if not integral with alternator)
Lubricating-oil filters (or elements)
Fuel-pump repair kit (or spare pump)
Spare glass bowl (if you have this type of water-trap/ pre-filter)
Propeller(s)
Gland packing for stern-tube
Miscellaneous nuts, bolts, coach-screws and washers

Bulbs for navigation lights, instrument panel and cabin lights
Insulating tape and plumber's tape
'Plastic Padding' or similar repair material
Solder and battery-operated iron
Lengths of insulated wire of different types and terminals
Piece of rubber sheet
Coil of copper wire
Length of stiff wire
Magnet (to be kept well away from the compass)
Asbestos lagging material.

For the diesel engine you will need:
Two injectors for a four-cylinder engine (and pro rata)
Two straight lengths of injection pipe with end-fittings to be bent to fit the particular cylinder
Fuel filter (or element)
Spare fuel-pipe olives or unions as used on your fuel system.

For the petrol engine you will need:
Sparking-plugs, ignition points, condenser, coil, distributor cap, spare high tension leads.

Tools

A good set of tools will help you to carry out repairs. Some engines need special tools to get at awkward nuts, for example, and these are usually provided in the kit available from the engine-supplier. A typical set of engine tools is shown in Figure 3.3. To deal effectively with installation fittings other tools will often be found useful – especially if you are working single-handed. I would add to the simple kit illustrated:

Vice grip pliers e.g. Mole self-grip wrench
Adjustable wrench
Chain wrench for gripping smooth surfaces, e.g. removing filter canisters
Set of socket spanners with universal joint, as well as various extensions
Wire-cutting pliers
Various screwdrivers
'Philips' head and 'posidrive' screwdrivers
Allen keys
Hacksaw – standard and 'junior' sizes

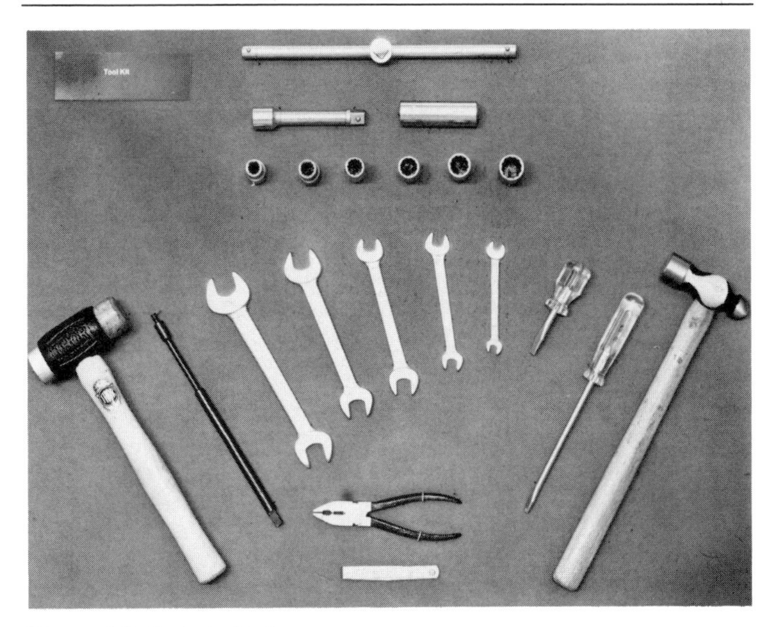

Figure 3.3 *Basic tool kit*

Propeller puller
Cold chisels
Feeler-gauges, ignition spanners, fine emery cloth.

Of course you will check that your spanners are suitable for the hexagons on the engine nuts and bolts. There are three standards for hexagon-headed nuts and bolts:

1. American SAE ('Across-flats' measurement in fractions of an inch)
2. Metric (Millimetre measurements)
3. British Standard used with BSF and Whitworth threaded nuts and bolts but now largely phased out. These hexagons are larger than the American SAE type for the same-sized bolt.

Some engines have an unfortunate mixture of hexagon sizes and in any case they may be different on the transmission or stern-gear and other fittings, so you will probably need types 1 and 2 as well as the Allen keys.

'Get-you-home' techniques

We have all heard tales of how the skipper sacrificed his new yellow wellies and braces to make a temporary repair to the rubber exhaust pipe, when a blockage of the water intake caused hot exhaust gas to burn the rubber hose. This is said to be a fairly common occurrence, although the person who listens to the note of the engine, checks the temperature gauge occasionally, and even sniffs the air over the transom now and then will detect such problems before they get out of hand. A spare length of diesel exhaust hose (which incorporates a reinforced steel spiral) is not easy to insert in the line, so it is more practical to wrap a length of rubber sheeting over the burnt section and secure it with hose clips. A slight seep of water from the repair won't matter, but don't forget to clear the blocked strainer or inlet fitting before running the engine again.

Splits in water hoses or cracked water pipes can be repaired by plumbers' tape.

A cracked header tank casting or water-jacketed exhaust manifold could leak water faster than you can top it up. If the crack is accessible a repair can be made with Araldite, Plastic Padding or similar 'two-pack' material. Dry off the surface to be repaired, clean and roughen before applying the mixture. Apply heat to accelerate the setting time.

This type of product can be used to plug holes if you lose a drain cock, but you must be able to clean the surface to be repaired – it may be easier to whittle down a piece of wood, then screw this into the hole with pliers or tap it in gently with the hammer.

An alternator or pump drive-belt can be replaced by multi-strand rope if you forget to bring a spare. Splice the ends of the rope together to make a continuous loop – if you can't make a proper splice do the best you can with the ends, finishing off by serving. Tension the alternator gently to avoid overstressing the rope ends, and run the engine slowly.

With any sort of luck you shouldn't have to deal with repairs like these very often – the main thing is to have all the right bits and pieces to hand so that you can avoid the ignominious prospect (and expense) of being towed in.

If you need to replace a gasket but don't have one available, a liquid such as Loctite Multi-gasket can be applied to the lower surface and will form a seal when the mating surfaces are brought together. It will also seal flanges and pipe threads.

Engine Breakdowns

Whatever the cause of an engine breakdown the result is always the same – the engine stops and you have got to get it going again with the minimum delay. You don't want to spend hours at anchor while you try to pinpoint the reason. A logical sequence of checks is required to establish the cause and then put it right, or at least make a temporary repair to get you home.

Petrol Engine Problems

A word of caution before you start work on the electrical or petrol supply systems: make sure that no one is smoking, the cooking gas is turned off and the bilge ventilation system is in operation.

Breakdowns affecting petrol engines are mainly caused by faults in the ignition system or in the fuel supply; the first thing is to decide which is the more likely cause. Spitting and popping from the carburettor indicates that insufficient fuel is getting through the jets; the blockage could be in the carburettor itself, the fuel-pump, piping, a filter screen or the tanks. It could be a vapour lock caused by overheated conditions in the engine compartment. A fuel shortage can also stop the engine quite suddenly, as can an ignition fault.

If you do not know whether the trouble is ignition or fuel, make sure that you have petrol in the tank, then check whether there is petrol in the carburettor float chamber – undo the bolts retaining the cover and see if it contains petrol. If it does, the blockage could be in the carburettor jets but this is unlikely – especially with the type of carburettor having a moving jet needle.

Now turn your attention to the ignition and check whether there is life in the high tension circuit by pulling the lead off a plug and exposing the terminal – this may necessitate unscrewing or pulling off a plastic terminal cover. Switch on and operate the starter while holding the plug terminal about 6 mm. (¼ in.) from the cylinder-head. If a fat spark shoots across the gap your trouble (if it is ignition) must be the plugs, although this is an unlikely cause of the engine's suddenly stopping. If you had stopped the engine and failed to restart, the plugs could well be the cause – oiled up or soaked in petrol because the carburettor choke was stuck.

If you obtained a good spark, try another lead for good measure, and, if this also worked, remove the plugs and fit your spare set. This is quicker than drying and cleaning the ones on the engine.

If you fail to get a spark from the plug leads, the fault is either elsewhere in the high tension circuit or in the low tension. Pull out the high tension wire from the centre of the distributor (the other end is attached to the coil) and hold this close to the cylinder-head while cranking the engine on the starter. If you see a good spark it is the distributor which is at fault. Remove the distributor cap and examine the rotor arm, check the spring contact in the centre, and make sure that the end of the lead is contacting the distributor terminal properly. If in doubt, fit your spare distributor cap – the fault could be a crack in the plastic that allows high voltage current to 'track' to earth.

If you did not get a spark from the distributor lead, the fault could be the 'points', condenser or wiring connection, i.e. the low tension circuit, or it could be the coil.

Check the points by flicking them open with a small screwdriver. You should see a small spark. If there is a fairly big one and the points look burnt it is probably the condenser which has let you down. Fit your spare condenser and clean up the points with the small file – just push this back and forth between the contacts; provided that they open and close properly you can leave the accurate setting-up until you are back on the mooring. If the points seem all right but you get no spark, check if there is current reaching the terminal on the outside of the distributor – connect your test lamp or voltmeter from the terminal to earth (or a negative connection) with the ignition switched on. If there is no current it is a wiring fault. Do not spend too long checking this unless you can spot a disconnected terminal or blown fuse. Fit a length of wire direct from the terminal to the battery positive connection or other source of current.

If there is still no spark, about the only thing left to check is the coil – you can only replace this by your spare one.

You may find the cause of an intermittent fault in the high tension wiring. The carbon filament used in the plastic-insulated wires in current use seem less reliable than the old rubber-insulated copper leads. After a few seasons' use you may experience misfiring – the only cure is to fit a new set of leads. Better still, change the leads after two or three seasons before trouble occurs.

If you found no petrol in the carburettor float chamber with plenty in the tank, either the supply pump is not working or you have a blockage in the pipeline or filters. Check the pre-filter with the tank shut-off valve on; fuel should leak out if the filter is below the fuel level. If not, slacken the union connection at the tank valve. If petrol runs out, retighten and work your way to the carburettor, cleaning the filter screens in the supply pump and carburettor intake. Fuel-pipe blockages can often be cleared by blowing down the section of pipe. Failing that, push a length of wire through the pipe.

If no blockage is found, check the supply pump. A mechanical type driven by the engine is the most common arrangement. Refer to the diagram on page 54. Check the condition of the diaphragm by removing the cover; replace if necessary. Electric pumps also have a diaphragm; they are more prone to failure of the electrics but will often respond to a tap on the casing. You might get home by administering a regular tap on this component. If you don't carry a spare and have checked that current is reaching the terminal and that the diaphragm is sound, remove the end-cover and clean the contact breaker points. You often get early-warning signs of electric lift pump failure with intermediate stopping and starting before it packs up altogether.

Let's now recap on the procedure:

1. Unless you are certain it is the ignition system, check if there is fuel in the carburettor float chamber.
2. With fuel in the carburettor, pull off a spark-plug lead and check for sparking against the cylinder-head.
3. If a spark appears, replace plugs.
4. With no spark from the plug leads, check the distributor cap, rotor arm, spring contact, points, condenser.
5. If there is still no spark, check the battery voltage supply to the distributor, coil, high tension leads.
6. If the petrol supply is at fault, check for a blockage in the fuel lines (blow through pipe), dirty filter screens, perforated fuel supply pump diaphragm, faulty electric lift pump (if fitted).

Diesel Engine Problems

One advantage of the diesel is that it does not have electrical ignition. Breakdowns of fuel injection systems are not unknown but are much rarer than ignition problems.

The main causes of failure are air or dirt in the fuel system. Air can be drawn into the fuel lines if connections are not properly made or if you run out of fuel and suck air instead of fuel from the tank. This can also happen in rough conditions when fuel swirls around in a tank which is not adequately baffled. Plastic pipe sometimes used for sections of the supply is not suitable. It can soften and work loose at the joints.

Small particles of dirt may bypass the pre-filter screen and the lift pump but should be trapped in the final filter if this has the correct paper element. Dirt passing this filter could do expensive damage to the fuel injection pump.

If the engine stops, check that all valves are fully 'on' and that you have fuel in the tank. Check the filter screen in the pre-filter and fuel lift pump; clean if necessary. Disconnect the flexible pipe at the engine inlet connection and blow through it. You should be able to blow into the tank. If this is blocked use a long piece of wire from the tool-kit if blowing does not unblock it. If your problem is not a blockage, check the fuel lift pump. Slacken the bleed connection on the fuel filter, then pump the priming lever. If the pump does not squirt fuel from the bleed connection, remove the pump cover and check the diaphragm; replace it if perforated. If the pump is difficult to work on in the boat, it may be easier to remove it and fit a replacement, in which case you should of course make provision in your 'on-board' spares kit.

The procedure for purging the fuel system of air varies from engine to engine. The principle is the same, however: work your way from the tank to the fuel-pump inlet, loosening the prescribed bleed screws and using the hand priming lever to expel air until no more bubbles are seen. It is often necessary to deal with the high pressure system after the low pressure plumbing is clear. This entails loosening an injection pipe connection at the injector inlet and motoring the engine over on the starter until air bubbles no longer appear in the leaking fuel. The engine will usually run on the other cylinders when the system is cleared, if it is more than a twin-cylinder; you then retighten the injector connection and continue running. (Figure 3.4)

Just as you have a warning system – by means of an indicator light and buzzer – to tell you the oil pressure is low or the water temperature is too high, you can install a similar system for low fuel level.

An elementary point here, but one that has been overlooked more than once: don't try to bleed the system without an adequate fuel level in the tank – you will just pump more air through the system until you realize what is wrong.

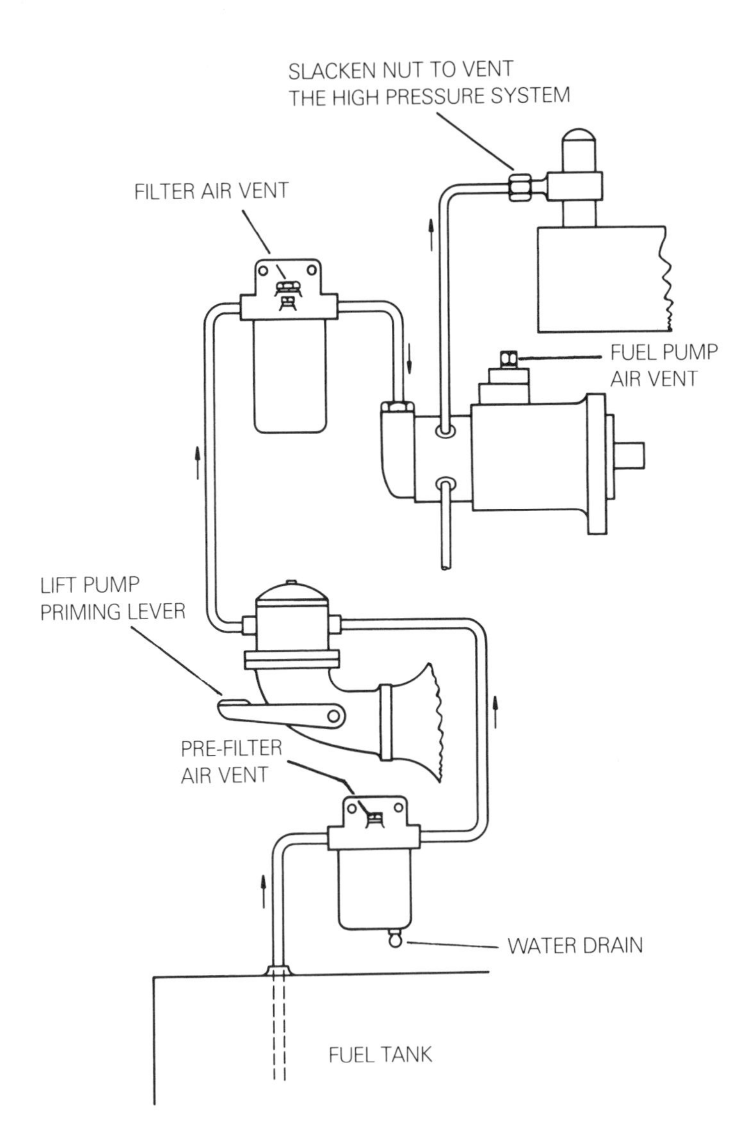

Figure 3.4 *Air-venting the diesel fuel system*

Diesel Fuel-pipe Leaks

Leaks in the pipe from tank to engine are not easily detected while the engine is still running because they cause air to be drawn into the system, eventually stopping the engine. With fuel in the pipeline when the engine is not running, it will leak out of the faulty connection or fractured pipe so that you can see where the problem is. Temporary repairs can be made to a fractured pipe by winding adhesive tape (the plastic type is best) tightly round the area, starting at least 25 mm. (1 in.) each side of the fracture. Plastic Padding or similar material can be used for temporary repair to an elbow or tank boss etc. You must clean the surface thoroughly and give the material the prescribed time to set. Warmth will accelerate the setting if you can arrange this safely.

When a high pressure pipe cracks and you don't have (or can't fit) a replacement, an engine of three or more cylinders can usually run without one of them functioning. Disconnect the broken pipe from the injector and lead it away so that the fuel is directed into a can or other convenient receptacle. Keep clear of the spray because the high pressure can easily penetrate your skin. (Figure 3.5)

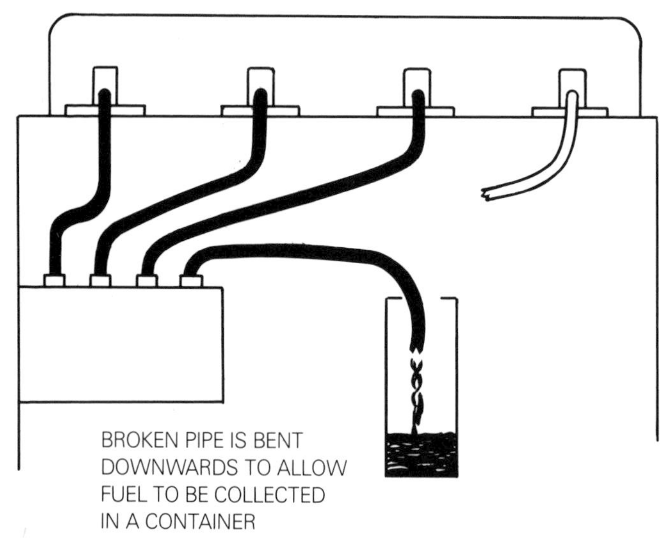

Figure 3.5 *Broken fuel injection pipe*

Diesel Fault-finding

Petrol engine faults can be diagnosed by electronic equipment which analyses the functioning of electrical and mechanical components, so that the service engineer knows which settings have to be restored to the specified level and where something is not performing correctly. Similar diagnostic equipment is available for diesel engines. Whereas the petrol engine diagnostics are largely concerned with the spark which ignites the charge in the cylinder, the diesel system displays a diagram of the *operating cycle*, i.e. what is happening in the cylinder through the two or four phases of the cycle. This diagram appears on a cathode ray oscilloscope, which is like a small television screen. The operator compares the display with a transfer showing the ideal diagram shape, and is able to decide what has got out of adjustment or where items are not functioning correctly. For example, a blip or series of blips at the end of the diagram means a secondary injection of fuel from the atomizer, which is wasting fuel and degrading the engine performance.

However, a lot of faults can be diagnosed simply by using your eyes and ears plus the handy chart developed by Perkins Engines, which is printed overleaf. Copies of this are available from Perkins engine-dealers and, as the information applies in principle to any make of diesel engine, it is a useful aid to trouble shooting that is worth keeping handy for use in emergencies.

Black Exhaust Smoke

This item on the list of faults has many possible causes. Some high-powered engines are prone to a dirty exhaust if the throttle is opened rapidly, particularly if the engine has been idling for some minutes. However, if black smoke and apparent loss of power occur at full throttle and the engine cannot attain its normal maximum speed, a faulty injector is a probable reason.

Carbon can build up in the small holes in the tip of the injector. The fuel is then not properly 'atomized', and, if the injector is tested on the bench, fuel will dribble from the hole instead of being emitted in a fine spray. The remedy is to prick out the hole with a nozzle drill; but some diesel engine manufacturers do not recommend this practice, so that it is best to fit your spare and return the faulty unit for stripping down, reconditioning and reassembly by the engine-distributor.

When refitting an injector be careful to tighten the two studs evenly so that the copper sealing washer makes a leakproof joint. It is best to use a new washer each time you replace an injector.

To determine which injector is faulty, slacken the pipe union nuts one at a time with the engine running at a fast idle speed. If the engine speed does not change with a particular injector bypassed, that is the faulty one. Be careful to avoid getting the spray on yourself.

Fault-finding chart

Fault	*Possible Cause*
Low cranking speed	1, 2, 3, 4.
Will not start	5, 6, 7, 8, 9, 10, 12, 13, 14, 15, 16, 17, 18, 19, 20, 22, 31, 32, 33.
Difficult starting	5, 7, 8, 9, 10, 11, 12, 13, 14, 15, 16, 18, 19, 20, 21, 22, 24, 29, 31, 32, 33.
Lack of power	8, 9, 10, 11, 12, 13, 14, 18, 19, 20, 21, 22, 23, 24, 25, 26, 27, 31, 32, 33.
Misfiring	8, 9, 10, 12, 13, 14, 16, 18, 19, 20, 25, 26, 28, 29, 30, 32.
Excessive fuel consumption	11, 13, 14, 16, 18, 19, 20, 22, 23, 24, 25, 27, 28, 29, 31, 32, 33.
Black exhaust	11, 13, 14, 16, 18, 19, 20, 22, 24, 25, 27, 28, 29, 31, 32, 33.
Blue/white exhaust	4, 16, 18, 19, 20, 25, 27, 31, 33, 34, 35, 45, 56.
Low oil pressure	4, 36, 37, 38, 39, 40, 42, 43, 44, 53, 58.
Knocking	9, 14, 16, 18, 19, 22, 26, 28, 29, 31, 33, 35, 36, 45, 46, 59.
Erratic running	7, 8, 9, 10, 11, 12, 13, 14, 16, 20, 21, 23, 26, 28, 29, 30, 33, 35, 45, 59.
Vibration	13, 14, 20, 23, 25, 26, 29, 30, 33, 45, 48, 49.
High oil pressure	4, 38, 41.
Overheating	11, 13, 14, 16, 18, 19, 24, 25, 45, 47, 50, 51, 52, 53, 54, 57.
Excessive crankcase pressure	25, 31, 33, 34, 45, 55.
Poor compression	11, 19, 25, 28, 29, 31, 32, 33, 34, 46, 59.
Start and stops	10, 11, 12.

1. Battery capacity low.
2. Bad electrical connections.
3. Faulty starter motor.
4. Incorrect grade of lubricating oil.
5. Low cranking speed.
6. Fuel tank empty.
7. Faulty stop control operation.
8. Blocked fuel feed pipe.
9. Faulty fuel lift pump.
10. Choked fuel filter.
11. Restriction in induction system.
12. Air in fuel system.
13. Faulty fuel injection pump.
14. Faulty atomizers or incorrect type.
15. Incorrect use of cold start equipment.
16. Faulty cold starting equipment.
17. Broken fuel injection pump drive.
18. Incorrect fuel pump timing.
19. Incorrect valve timing.
20. Poor compression.
21. Blocked fuel tank vent.
22. Incorrect type or grade of fuel.
23. Sticking throttle or restricted movement.
24. Exhaust pipe restriction.
25. Cylinder-head gasket leaking.
26. Overheating.
27. Cold running.
28. Incorrect tappet adjustment.
29. Sticking valves.
30. Incorrect high pressure pipes.
31. Worn cylinder bores.
32. Pitted valves and seats.
33. Broken, worn or sticking ring(s).
34. Worn valve stems and guides.
35. Overfull air cleaner or use of incorrect grade of oil.
36. Worn or damaged bearings.
37. Insufficient oil in sump.
38. Inaccurate gauge.
39. Oil pump worn.
40. Pressure relief valve sticking open.

41. Pressure relief valve sticking closed.
42. Broken relief valve spring.
43. Faulty suction pipe.
44. Choked oil filter.
45. Piston seizure/pick up.
46. Incorrect piston height.
47. Sea-cock strainer or heat-exchanger blocked.
48. Faulty engine mounting (Housing).
49. Incorrectly aligned flywheel housing, or flywheel.
50. Faulty thermostat.
51. Restriction in water-jacket.
52. Loose water pump drive belt.
53. Gearbox or engine oil cooler choked.
54. Faulty water pump.
55. Choked breather pipe.
56. Damaged valve stem oil deflectors (if fitted).
57. Coolant level too low.
58. Blocked sump strainer.
59. Broken valve spring.

Correcting Faults

Some of the fifty-nine diesel problems listed, as well as ones affecting petrol engines, are most likely to occur when you start up the engine after a few weeks out of service. If you neglected the laying-up procedure at the end of the season they are even more likely to occur.

Sticking valves can occur on both petrol and diesel engines – particularly when carbon forms on the valve stem and causes it to jam in the guide. It is easy to check if valves are sticking by watching which ones do not close when you motor the engine over on the starter, with the cylinder-head cover or side cover removed, according to whether the engine has overhead or side valves. If you have a starting handle you can turn the engine slowly with less risk of spraying yourself with oil than when motoring it with the starter.

The pistons usually hit sticking valves on an overhead valve engine, and a squirt of penetrating oil down the guide will often get the valve moving freely again. If not, remove the cap and spring (see Chapter Five) and work the valve up and down in the guide with more penetrating oil until it moves freely. Incidentally, do this with the piston less than half way down the bore; otherwise the freed-off valve can slip from your grasp and you will have to remove the cylinder-head to retrieve it! With side valves – more common on older petrol engines, which are also more susceptible to carbon build-up on the valves – the problem is the same but access is often more difficult than for overhead valves.

Stuck inlet valves can sometimes be eased by squirting penetrating oil into the air intake. Too much penetrating oil is not good for the bearings, so if you have used more than a few squirts you should drain the sump and refill with clean oil.

Carbon build-up can be minimized by avoiding prolonged running at low throttle settings with a very cold engine – particularly for the diesel. It is a good idea to run at full engine speed for a minute or so before shutting down, and this will also help to avoid blocked diesel injectors.

Petrol Engine Fault-finding

The most common petrol engine faults are given below, with possible causes:

Fault	*Possible Cause*
Engine does not fire	Distributor cap dirty or cracked Carbon brush worn or stuck Faulty plug or coil leads Faulty low tension wires Corroded terminals Dampness Distributor rotor cracked Faulty coil Points not opening Faulty condenser Spark-plugs oiled up or petrol soaked Petrol supply or carburettor problems
Engine misfires	Dirty spark-plugs Incorrect plug gaps Cracked plug insulator Loose plug(s) Incorrect ignition timing
Engine lacks power	Ignition timing incorrect Centrifugal advance flyweights seized Centrifugal advance springs weak Worn-out spark-plugs Distributor loose in housing
Engine stalls	Ignition system or fuel problems (see under 'Engine does not fire/ misfires') Mixture too weak Water in petrol Incorrect valve clearances

Fault	*Possible Cause*
Unsatisfactory idling	Air-leaks into manifold joints Carburettor wrongly adjusted Carburettor air-leaks Mixture too rich Engine needs top-overhaul
Engine vibrates	Mounting brackets or rubber mounts loose Water-pump or alternator loose Misfiring
Overheating	Thermostat not opening fully Blocked pipes or water passages Sea-cock or strainer turned off or blocked Collapsed or perished hoses Water-pump drive or belts faulty Low fresh-water level in header tank Blockage in exhaust system Weak petrol mixture Incorrect ignition timing Incorrect valve timing
Cooling water leakage into lubricating-oil	'Blown' cylinder-head gasket Cracked cylinder-head Loose cylinder-head securing nuts/screws Oil-cooler tubes or joints leaking

4 Tuning

We tend to think of tuning in connection with high-powered engines for racing cars, motorcycles and boats; indeed, maximum performance – whether of the boat itself or its engine – is obtained by tuning. However, we are not always interested just in performance: economical and reliable operation is also obtained by engine-tuning.

So what is tuning? Essentially it consists of restoring the various settings, clearances, timing and other variables in the functioning of the engine to the conditions which will 'optimize' its performance. These conditions are specified in the workshop manual – at least most of them are although some, such as 'dwell', may not be quoted. (More about this later.)

There is little point in tuning an engine that is in poor mechanical condition. If your sparking-plugs are becoming oiled up because the piston rings are worn and lubricating-oil is finding its way to the tops of the cylinders, a brand-new set of plugs will be wasted. So too will be the rest of the tune-up operation, so start with the engine in reasonable condition, which may entail some work as described in other chapters.

When the engine was being designed, settings such as the timing of the inlet and exhaust valve opening, the ignition timing (or fuel injection if we are considering a diesel engine) and many other variables will have been decided arbitrarily by the designer. The development engineer running the prototype engine on the test bed will have experimented with all these settings and most probably will have changed some of them. In some cases he will recommend alternatives for maximum performance or economy. These alternatives will not generally be quoted in the manual and a compromise will have been decided, unless of course the engine is for high performance operation, when economy tends to go out of the window and a high compression ratio, twin carburettors and perhaps turbo-charging will be resorted to.

The diesel is not quite so easily 'suped up' as the petrol engine. Ways of getting more air into the cylinders to burn more fuel are not easily achieved except by means of turbo-charging, although there is one method known as 'ram-tuning' of the air induction which improves performance.

So let's return to the main objective of tuning – getting all the variable settings adjusted to the conditions which the development engineer found to be optimum for your engine and so recorded as 'standard' in the workshop manual.

Tuning – like any other job on the engine or on the boat itself – may require new parts, and you won't know what is needed until you are well into the job. If you are lucky enough to have your mooring near your local engine-dealer, you can make up your shopping list as the work proceeds. Otherwise you should get hold of the items most likely to be needed in advance. These will generally be found in the 'on-board' spares kit which you can therefore rob, but you should of course replace them before going to sea.

Professionals usually take an hour or so to tune a petrol or diesel engine – you should manage it in about twice that time if you have the necessary tools and a portable vice.

Tuning the Inboard Petrol Engine

Although it is possible to do a fair job with a set of ignition spanners, feeler-gauges, pliers, screwdrivers, a wire brush and the workshop manual, a better result is obtainable with the aid of electronic tuning equipment. You don't need the expensive and bulky items used by garages – car accessory dealers will sell you a timing light, a mixture-indicating device and a dwell-meter for little more than the cost of one professional tune-up. If you don't want to buy these for a once-a-season job you will probably be able to borrow them from a friend or colleague who uses them on his car or motorcycle engine. (Plate 4.1)

Plugs

Start with the sparking-plugs. We have already referred to these under Preventive Maintenance, and it doesn't do to disturb them

Plate 4.1 *Tuning equipment for petrol engine (Gunson)*

unnecessarily; but if they have not been checked for some months, then remove them and the leads one at a time so that you don't get the leads crossed when refitting (although you will soon find out if you have made a mistake when you start up). Change the plugs if they are more than one season old; if their condition is good, set the gap by adjusting the clearance between the two electrodes, blow out the carbon, clean up with a small wire brush and refit. (Figure 4.2)

Note the appearance of the electrodes – whether sooty, white or whatever (see page 55). The gap is usually specified between 0.025 and 0.028 in. (0.60 to 0.70 mm.) Don't bend the centre electrode

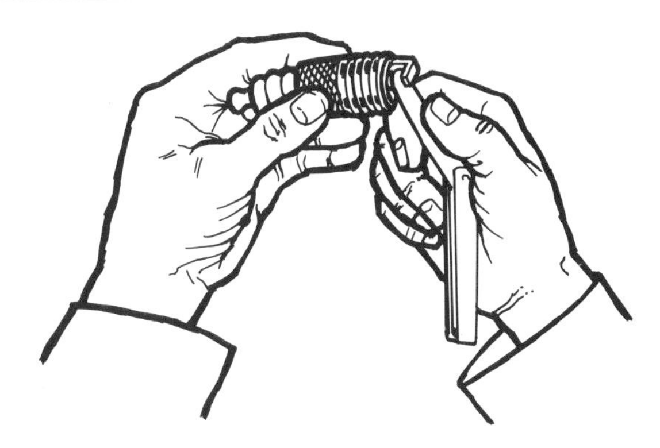

Figure 4.1 *Sparking-plugs. Setting the gap with a feeler gauge.*

as you could damage the insulation – always adjust the side electrode. Don't forget to check the gaps on new plugs and make sure they are of the correct type for your engine. Even if they have the correct 'reach', i.e. length, different grades of plug are specified by the manufacturers. If the washers seem to have been squashed flat you should replace them by new ones, and don't overtighten the plugs when refitting.

Leads

As you disconnect each lead from the sparking-plug check that the terminal fits tightly and is free from corrosion. Clean up the contacts with emery-paper or sandpaper. Wipe the leads with a clean cloth before examining the insulation for cracks. Next remove the distributor end and check the terminals. If you are not satisfied with any of the leads or they are more than three years old, play safe and replace the lot. Many of the failures occurring in the electrical system are caused by faulty leads, generally where there is a small break in the conductor or a fault in the insulation, causing the high tension current to jump across to earth instead of producing a spark at the plug points.

Distributor

Before removing the cap, thoroughly clean the outside – taking care not to damage the wire entering the side of the distributor. Spring back the clips and remove the distributor cap. Wipe out the inside and scrape off any corrosion from the brass contacts. Pull out the spring-loaded centre connector a little. If there is any damage to the plastic or signs of excessive wear or corrosion on the contacts, replace by a new cap. Very small cracks can develop in time and cause short-circuiting of the high tension current. This bypasses the plugs, causing problems similar to faulty plug leads.

Next pull off the rotor arm and check this for wear. If it looks well-worn, replace it. You may think this policy of 'if in doubt, replace it' a little extravagant, but what we are doing is insuring against a failure in a few months' time when you may be in a critical situation.

With the rotor arm out of the way, have a good look at the points which open and close by means of the rotating distributor cam. Turn the engine over to check the action. This is easy if you have a hand-starting engine. If not, pull the water pump/alternator driving-belt. A further method if you have a large spanner or adjustable wrench is to engage the nut which usually retains the crankshaft pulley. Yet again if you have a mechanically-operated gearbox, wedge a large screwdriver under the nuts on the propshaft flange – with the gear lever in 'ahead' or 'astern' of course. Removing the plugs makes turning easier.

Now turn your attention to the points. One is fixed and the other moves with the cam, forming the all-important gap with the 'make and break' action that produces the high voltage current in the ignition coil. Wedge the points open as far as possible with a screwdriver and inspect the contacting surfaces. If these are not badly worn or burnt you can clean them up. A small file is sold for this purpose, or alternatively you can wrap a piece of fine emery-paper or sandpaper around a thin strip of metal. If your by now high inspection standards are not met, fit a new points set. Have a good look at the position of the various components before removing the old points and refitting.

Next you have to set the gap, i.e. clearance, when the cam opens the points to the fullest extent. With the engine turned so that one of the cam lobes is bearing directly on the contacting surface, adjust the gap by slipping the appropriate feeler-gauge between the points. You may have to use two blades to make up the specified gap, which is generally 0.015 in. (0.38 mm.) – Figure 4.2. If you are using the original points and the gap is about one 'thou' (0.001 in. or 0.025 mm.) of the specified value, that will be near enough. Again, if using the original points make sure that the moving arm is free on its pivot. If not, polish lightly with a strip of fine emery-cloth and apply a drop of oil after removing any residual emery dust. (Figure 4.3)

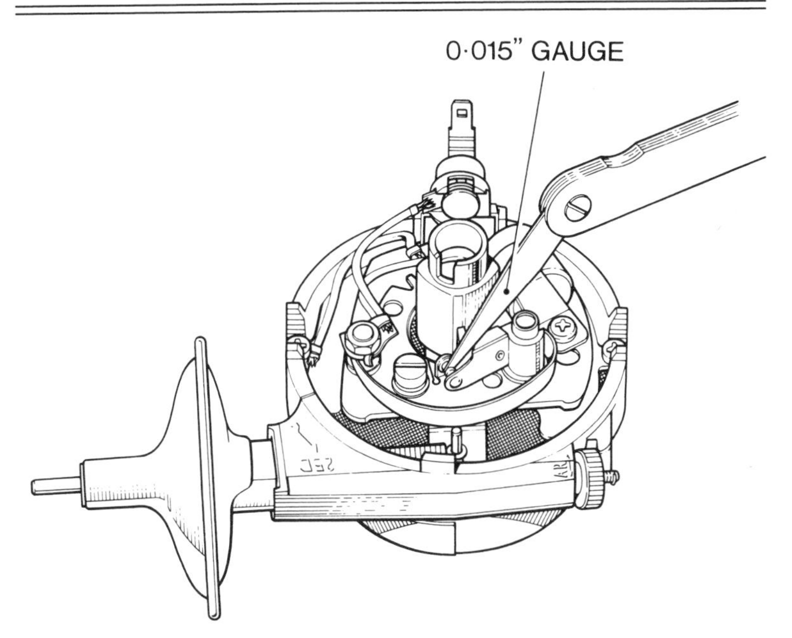

Figure 4.2 *Setting the contact points gap (Lucas)*

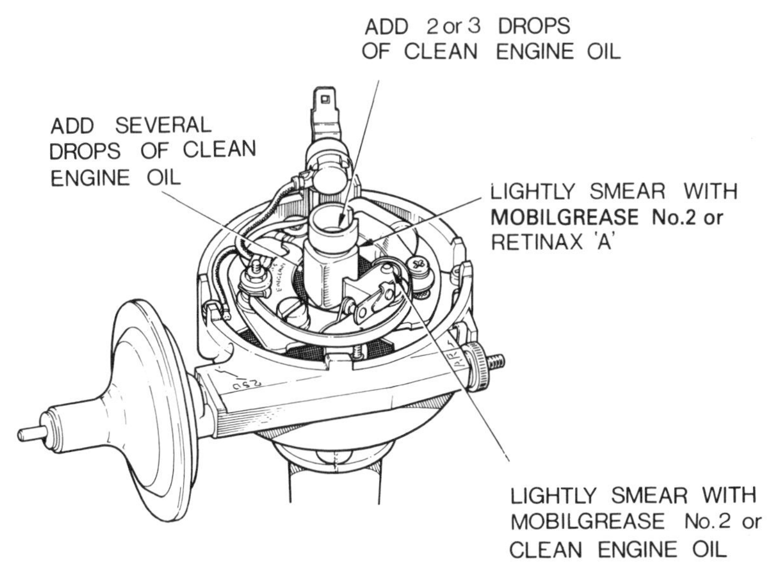

Figure 4.3 *Lubricating the distributor (Lucas)*

The centrifugal advance mechanism is located in the distributor body below the plate to which the points and the capacitor are fitted. Trouble with this device is not usual unless rust has caused the flyweights to stick. The operation of this and the vacuum advance can be checked later, so for the moment replace the cap after having squirted some thin oil into the mechanism, making certain you avoid oiling the points at the same time.

Ignition Coil

Without special test equipment you can only check the general condition of the wiring and connectors – also corrosion on the casing and retaining clips. The professional tuner would check whether the voltage produced by the secondary circuit – i.e. the output feeding the distributor – is up to standard when the starter is turning the engine over and at operating speeds. Over 20,000 volts would be obtained from a satisfactory coil.

Capacitor

There is nothing you can do to check the efficiency of this component without using special equipment. It only merits attention if you have checked everything else and suspect that this may be causing a failure of the ignition system; if so, your only option is to replace it by a known good unit.

Ignition Timing

When the engine is being overhauled, the standard timing point – No. 1 piston at T.D.C. (Top Dead Centre) – can be observed with the cylinder-head removed. In order to time the engine without removing the cylinder-head, an external indication of top dead centre is provided, also markings allowing you to check angular positions before the piston reaches the top of the cylinder.

Timing marks are often found on the crankshaft pulley with a pointer attached to the front of the engine. On some engines the marks are etched on the flywheel, and when this is enclosed an inspection cover is provided. (Figure 4.4)

As explained in Chapter One, the ignition timing varies at different engine speeds. It can also be measured in different ways. The 'initial' timing, i.e. the position of No. 1 piston in the cylinder when the sparking-plug fires, can be measured with the engine stationary or when it is running slowly before the mechanical flyweights or the vacuum advance device have any effect. The

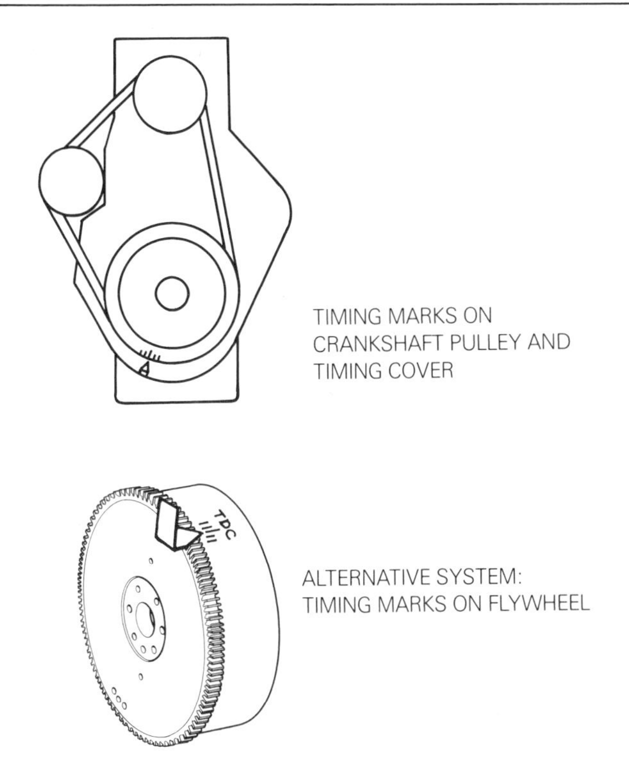

Figure 4.4 *Timing marks*

advantage of measuring the initial timing with the engine running is that the clearances in the gears or chain driving the distributor are taken up and you are measuring what is actually happening.

To gauge the timing with the engine stationary you have to judge the position at which the points are just opening. This can be done by placing a very thin piece of plastic film between the distributor points and carefully inching the engine over until you can pull it out. A more accurate method is to connect a lamp of the same voltage as the engine battery, across the points. When the lamp lights, the points have opened and current is flowing to the coil. If you have not disturbed the distributor position you should be very close to the static timing setting, which is generally T.D.C. Refer to the handbook or workshop manual for confirmation, however.

If the points are adjusted with too much clearance, they will start to open late and the timing will be retarded. Conversely, too little clearance will give advanced timing.

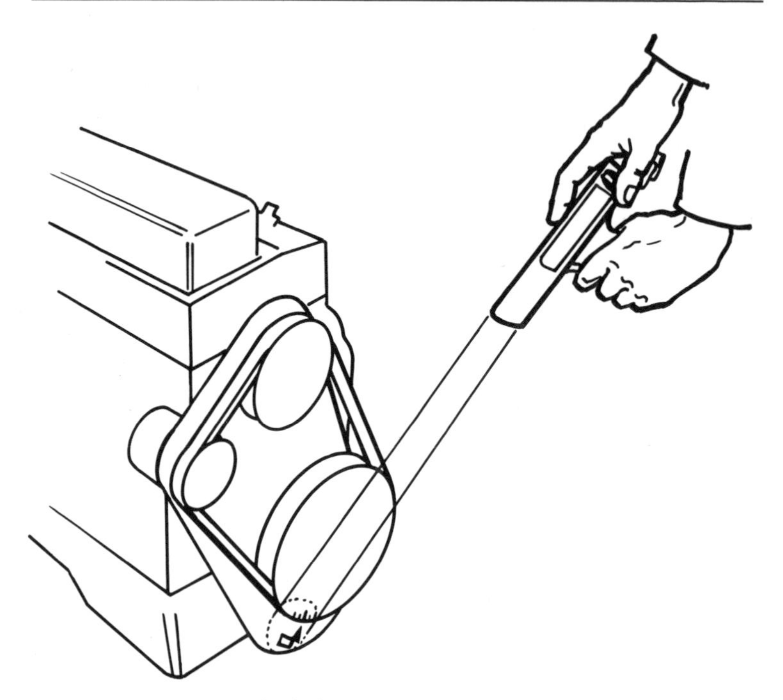

Figure 4.5 *Stroboscopic timing*

To check the ignition timing dynamically, i.e. when the engine is running, you will need a timing light or 'strobe'. This is a stroboscopic light wired to the ignition circuit so that it flashes each time No. 1 plug fires. By aiming the light at the timing mark this will appear stationary, and you can see whether the pointer and mark coincide or whether adjustment of the distributor clamp is needed to change its angular position in the support, and thus either advance or retard the timing – Figure 4.5. Moving the distributor in the direction of rotation of the rotor arm will retard the timing, and moving against the direction will advance it. Some distributors have a screw adjustment device allowing for very small movement, others have a clamp bolt which has to be slackened off so that the distributor can be turned bodily.

Incidentally, some of the stroboscopic timing lights are connected to the battery to intensify the light emitted – useful if you are working in sunny conditions. Another type has to be plugged into the mains supply – not much use to you unless you are moored at a marina with the necessary electricity laid on.

Another advantage of checking ignition timing dynamically – sometimes known as 'power' timing – is that you can check whether the automatic advance devices are operating. As the engine speed is increased the opening of the flyweights in the distributor advances the timing; you can see if this is happening because the timing pointer will appear to move away from the T.D.C. mark.

With the aid of a tachometer to measure engine speed accurately you can check whether the advance is working correctly to specification, which means starting to open at the prescribed speed and moving for so many degrees during the period when a higher specified speed is attained. However, this is perhaps being too precise for a DIY tune-up – it will suffice if the device is operating at all, and this can be clearly demonstrated with the 'strobe'.

Dwell

This is the period – expressed in degrees – during which the points remain closed and the battery power is flowing through the primary windings of the coil, causing the spark to be produced at the plug. The duration of the dwell is important to the quality of the spark produced, and it is dependent on the clearance between the points and the condition of the cam lobes. If you set the points too wide you get a small dwell angle, and conversely a narrow gap gives a larger dwell. Because of variations in the shape of the cam (caused largely by wear) the dwell angle can vary, and it is recommended that you measure this if you want to obtain maximum operating efficiency.

A dwell-meter can be connected up to the low voltage circuit in the manner prescribed in the instructions provided with the instrument, and the dwell is read off directly on the meter. Some meters also give you a reading of the voltage at the coil and the condition of the points, so that you can check these before taking the distributor cover off, although you should check it over visually as described earlier.

You may not find dwell quoted in the workshop manual for your engine. It is usually between 38 and 48 degrees for four-cylinder engines and less for six- and eight-cylinder ones. If you do not have the figure, check with your dealer when you are buying the points and plugs etc.

If you have set the clearance within the specification and the dwell is not correct, you will have to readjust the clearance slightly outside specification until the dwell is corrected, because this is the more important factor. (Figure 4.6)

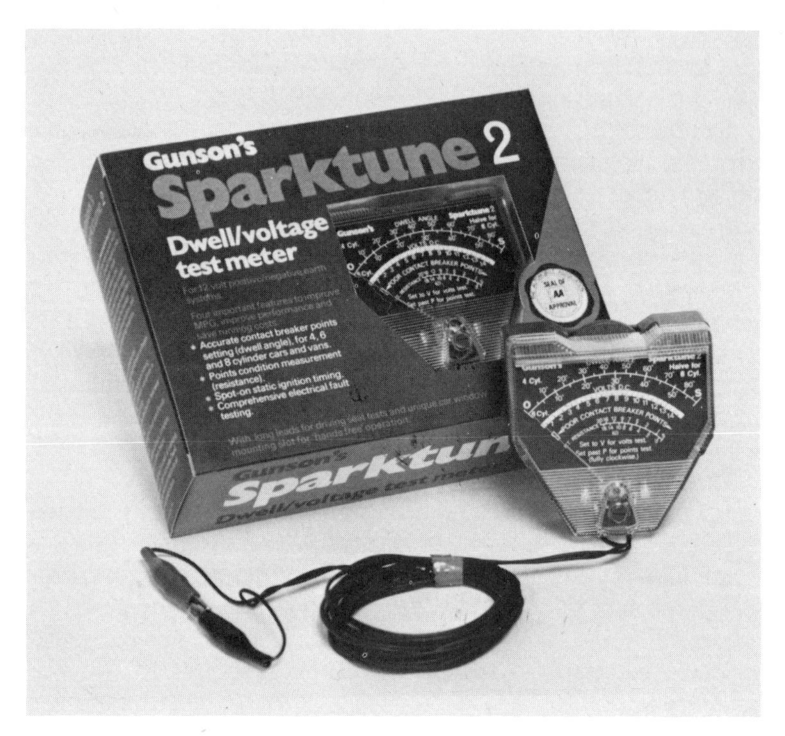

Figure 4.6 *Dwell meter (Gunson)*

Valve Timing

It is unlikely that wear of the timing chain or gears – whichever is used in your engine – will cause the opening and closing of the inlet and exhaust valves to vary sufficiently to affect engine performance. If the valves do not seal properly because of carbon on the seats or weak springs, this is a different matter and the cylinder-head probably needs decarbonizing (a process which is described in Chapter Five).

You can check the effectiveness of the valves by measuring compression pressure with the engine running – see Figure 4.7. The likelihood of a problem in the first few hundred hours of engine life is not great, and an engine which idles slowly and runs smoothly will generally be satisfactory as regards valve setting. However the tappet clearances are likely to be outside the figures quoted in the manual or indicated on an instruction plate fixed to the engine. A few 'thous' of extra clearance means several degrees of retardation

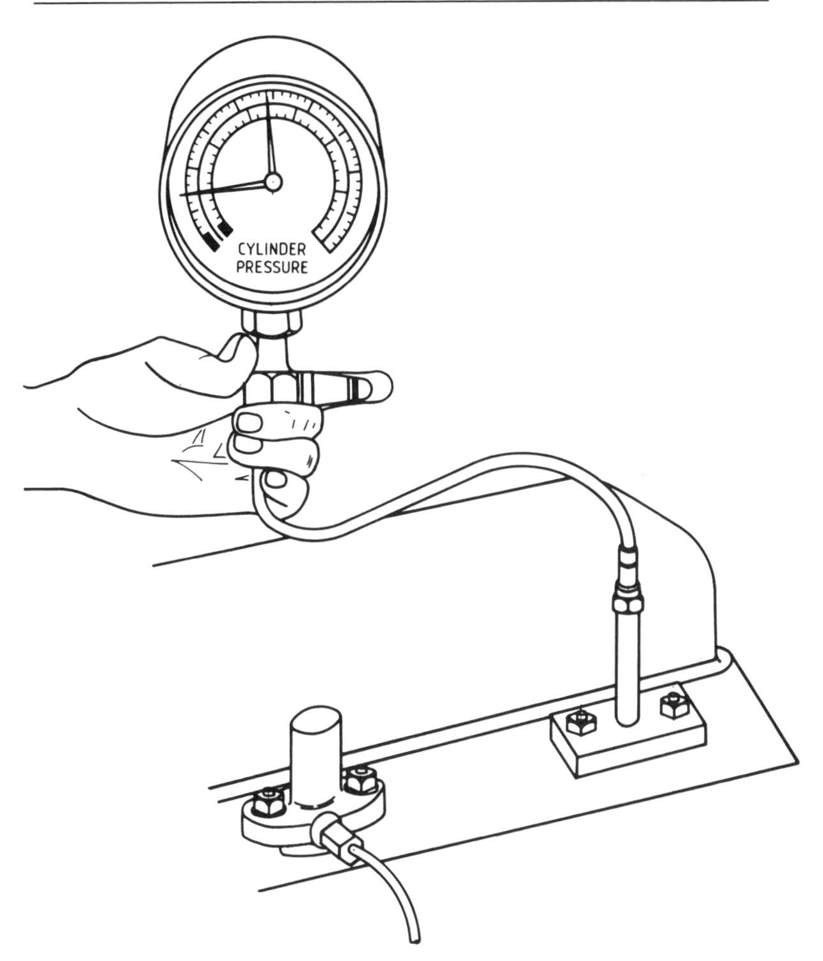

Figure 4.7 *Measuring compression pressure. The illustration shows an adapter flange replacing a diesel fuel injector. A different adapter replaces a sparking plug for petrol engines.*

in the points where the valves start to open and close, so they should be carefully checked with the feeler-gauges and reset if necessary.

Carburettor

It is important to tune the carburettor after all the other components have been dealt with. You could adjust the fuel mixture and then find that correcting the ignition timing, for example, made the idling unsatisfactory and necessitated readjustment of the carburettor.

As with the ignition system, considerable experimentation will have taken place during the development of the engine to arrive at the best carburettor settings. After many hours of running, wear will take place in the jets, screws may have vibrated loose and there is a need to restore conditions to 'standard'. Improvements in power or economy may be possible by changing jets or replacing the carburettor by a special model, but it is often necessary to compliment carburettor modifications with matching changes to plugs or the ignition system; and, to reap the benefit, the engine needs setting up by an expert. We will concentrate therefore on getting the carburettor back to where it was (or should have been) when the engine was installed.

As explained in Chapter One there are two main categories of carburettor, although many different makes and models are produced by manufacturers throughout the world. Workshop manuals give detailed instructions for adjustment, but we will go through the procedure for one of each type.

Whatever model is fitted to your engine, the aim is to have an air/fuel mixture which gives economical running without power loss throughout the operating speed range. Some clues to whether the engine is operating with an approximately correct or definitely incorrect mixture can be obtained by examining the condition of the sparking plug electrodes. A sooty appearance denotes an over-rich mixture, while a white colour indicates weak mixture conditions. In ideal conditions browny grey powdery deposits are present – if you have these your carburettor setting is not likely to be far out. Don't be tempted to run the engine with too weak a mixture. Apart from a tendency to pop back through the carburettor and a reluctance to build up speed, the engine will run too hot and you may get trouble with burned-out valves.

Start your carburettor tune-up by cleaning the outside; lubricate the throttle and choke controls and check the attachment flange screws for tightness. Next, start the engine and warm up. You will reach working temperature more quickly if you can put some load on the engine by engaging gear. Run for five minutes after working temperature is reached, then give a quick burst of throttle (in neutral). Get on with the tuning before the engine temperature drops. If more than three minutes elapses you must warm up again.

The foregoing applies to any type of carburettor, but we will start with an SU type which has a variable jet and is illustrated in Figure 4.8.

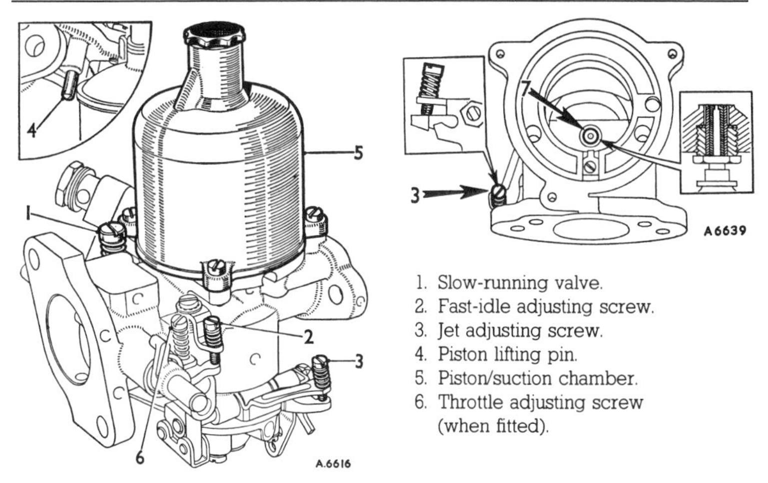

Figure 4.8 *Tuning SU carburretor type HD*

Unscrew the fast-idle adjusting screw (2) to clear the throttle stop with throttle closed; screw down the slow-running valve (1) onto its seating, then unscrew $3\frac{1}{2}$ turns. Remove the piston/suction chamber unit then turn the jet adjusting screw (3) until the jet (7) is flush with the bridge of the carburettor. Replace the piston/suction chamber unit and check that the piston falls with a click when the lifting pin (4) is released – if not the jet needs to be re-centred – best done by an experienced mechanic. Lower the jet by turning the adjusting screw (3) down $2\frac{1}{2}$ turns.

Restart the engine and adjust the slow-running valve (1) to give the desired idling speed. Turn the jet adjusting screw (3) up to weaken or down to enrich until the fastest idling speed with even running is obtained. Re-adjust the slow-running valve (1) if necessary to give correct idling speed.

Check for correct mixture by gently pushing the lifting pin (4) up about $\frac{1}{32}$ in. (1 mm.) after free movement has been taken up. If the mixture is too rich the speed will rise considerably; if correct, the speed will rise slightly, but if weak it will immediately decrease. Re-adjust the mixture strength if necessary.

Re-connect the choke control cable allowing about $\frac{1}{16}$ in. (1.5 mm.) free movement at the knob before the jet is moved. Adjust the fast-idle screw on the choke linkage to give about 1000 r.p.m. engine speed, i.e. with the ignition light out when hot. Finally, top-up the piston damper with the recommended engine oil until the level is $\frac{1}{2}$ in. (13 mm.) below the top of the hollow piston rod.

With multiple-SU carburettors it is necessary to balance them as well as setting each one up as described for the single model. Use a short piece of PVC tubing like a stethoscope to listen to the air intake 'hiss', adjusting the mixture strength until these sound the same. The interconnecting linkage is disconnected before tuning and afterwards it has to be carefully reconnected so that the carburettor controls work in unison. Air flow into twin carburettors can be measured by using a device obtainable from car accessory dealers, which is fitted to each carburettor and has a scale recording the air flow.

Watermota/Ford Carburettor

This is a single choke, downdraft carburettor with an accelerator pump. Two adjustments are provided – a slow running speed (throttle stop) screw and a volume control screw, for mixture adjustment. (Figure 4.9).

After warming up the engine, the slow running screw is adjusted to give a speed of approximately 800 r.p.m. If you do not have a tachometer, this can be measured with the stroboscopic light as used for dynamic timing. See page 94. Next turn the volume control screw anti-clockwise until the engine 'hunts', i.e. runs irregularly. Now carefully and slowly turn the control screw clockwise until the engine runs evenly. Re-adjust the slow running screw to give 800 r.p.m.

Note that a brand new engine that has not been run in will not idle perfectly at low speeds.

Tuning the Diesel

The major difference between a petrol and a diesel engine is that the former has a spark ignition system while the latter ignites its fuel-and-air mixture without the need for a spark, relying on a much higher compression pressure in the cylinder to generate the necessary temperature for combustion. The main feature of the petrol engine requiring tuning has thus been eliminated.

Although diesel engines will run for many hundreds of hours without the need for a tune-up they do eventually require attention. While the work-boat engine could need an FIE overhaul every year, a sailing-boat auxiliary is likely to run many seasons and then only need attention because of a problem arising during a long winter lay-up. Perhaps it is just as well that, given the necessary

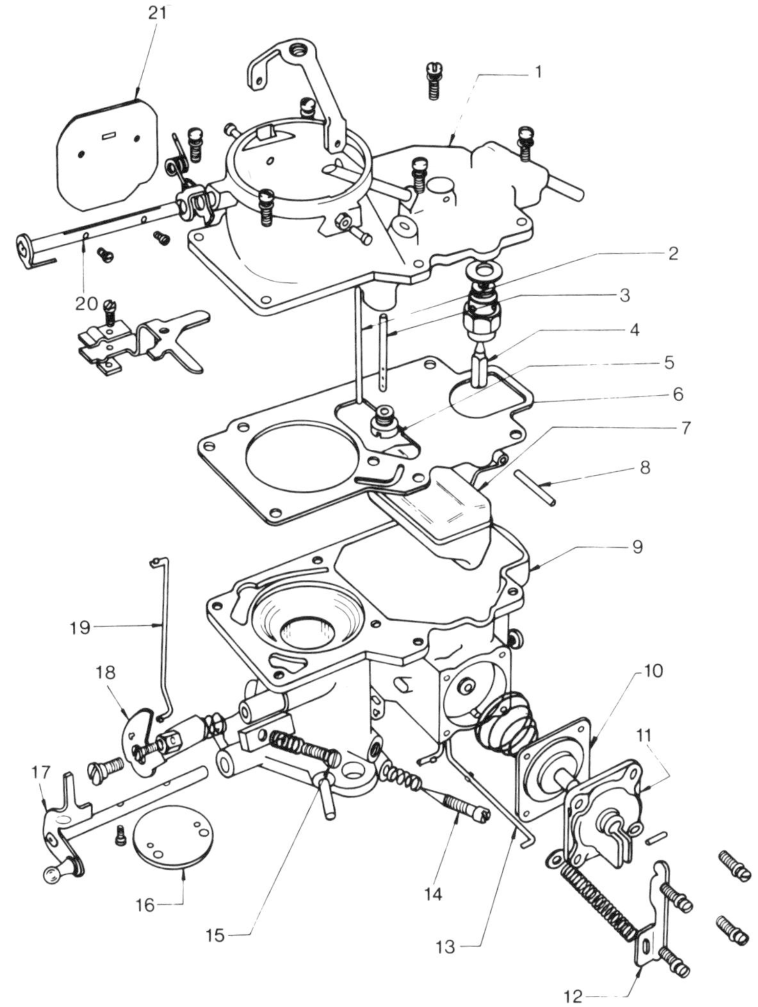

Figure 4.9 *Watermota – Ford carburettor*

1 Float chamber cover and upper body
2 Dip tube
3 Emulsion tube
4 Needle valve
5 Main jet
6 Gasket
7 Float
8 Float pivot pin
9 Carburettor body – lower
10 Diaphragm
11 Accelerator pump
12 Operating lever
13 Accelerator pump rod
14 Volume control screw
15 Throttle stop screw
16 Throttle plate
17 Throttle control lever and spindle
18 Fast idle cam
19 Choke link
20 Choke spindle
21 Choke plate

preventive maintenance, diesels in pleasure-craft will go many seasons without tuning, as this procedure is not easy for the DIY man who is quite capable of dealing with his inboard or outboard petrol engine. Nevertheless we will look at the tuning requirements of the diesel.

Fuel-injection Pump

This very reliable piece of equipment is designed to inject a minute, precisely metered drop of fuel into the cylinder, varying the quantity according to the load on the engine and, on a multi-cylinder model, distributing the same amount to each cylinder so that no cylinder has a 'richer' or 'leaner' mixture than its fellows.

After countless millions of injections the mechanical parts of the pump will start to wear a little, although maintaining a higher standard of fuel filtration will prevent abrasive particles from accelerating the wear process. With wear to the plungers and the mechanism controlling the displacement of these pumping elements, the amount of fuel metered will vary; usually it will increase so that more fuel is consumed and the exhaust becomes smokier. The timing of the fuel injection may be affected by wear, and the pump governor can also change because of wear in the flyweights, control linkages or other parts. The only practicable answer is to have the injection pump overhauled by a service dealer. You can do this via the engine-dealer, who may have the necessary specialized equipment, or the fuel-injection pump manufacturer's own dealers. CAV and Bosch, for example, have their dealer networks in most countries.

It is easy to remove the pump and replace it on the engine, as there will be timing marks that vary with the make of engine, but which are similar in principle to those used with the distributor of a petrol engine. The workshop manual will explain the timing procedure.

Injectors

Before considering having the fuel-injector pump overhauled, make sure that the injectors are not faulty. These will require more frequent attention because of the build-up of carbon in the tip which projects into the cylinder. Chapter Three explains how to detect a faulty fuel injector and replace this by a spare. Though it is a fairly simple matter to strip an injector assembly and fit a new nozzle, to set it up to the correct spring pressure requires the use of a bench test assembly which it is scarcely worth purchasing unless you run a fleet of boats. The spare set of injectors therefore should be kept on board for replacement purposes, and the faulty ones returned to the dealer for overhaul as necessary.

The Drive Train

Having brought all the engine adjustments into line with the manufacturer's recommended settings to assure economical and reliable operation, you should make sure that the transmission system, i.e. gearbox, propeller-shaft and bearings, are in good condition and not absorbing power unnecessarily.

A tight bearing in the gearbox or stern-tube, a bent shaft, or a damaged or corroded propeller can waste the power that your engine tune-up restored.

Disconnect the propeller-shaft coupling behind the gearbox flange, slide the shaft back until the flanges are separated, then turn the shaft to see if the faces of the flanges are parallel, to check whether the shaft is bent. Correcting a bent shaft will require removal from the boat and is a job for the workshop, where the shaft can be placed in vee-blocks to check its straightness.

Worn bearings in the stern-tube should be replaced. A damaged or corroded propeller can be restored to new condition by a propeller specialist.

If the propeller-shaft is out of alignment with the gearbox flange, the extra force required to rotate the shaft will waste power as well as possibly damaging bearings in the gearbox or stern-tube. Check the alignment with the boat in the water, as described on page 70, and if necessary adjust the engine mountings to level up the engine.

If you still do not achieve the desired engine speed after the engine tuning and possibly work on the drive train, you may have an unsuitable prop (i.e. diameter), pitch or number of blades. Before trying alternative propellers it is best to obtain expert advice from a specialist. It may well be that your boat would not go any faster even with the higher engine speed you could get with a smaller propeller. Many displacement hulls do not go faster with more power; the engine simply makes more noise and uses more fuel.

Of course a deterioration in boat speed is likely if the bottom needs scrubbing and repainting. An increase in displacement caused by additional weight of fuel, stores and passengers will have the same effect – particularly with a planing hull.

5 Engine repair and overhaul

A regular maintenance schedule, with the replacement as necessary of external items such as water-pumps, thermostat, alternator, starter and, of course, petrol engine ignition components or diesel engine fuel-injection parts, will generally ensure many years of service before the internal parts of the engine need attention.

However, typical operating conditions in a petrol- or diesel-engined cruising boat, i.e. part throttle running with the engine over-cooled, will speed up the formation of carbon in the cylinder-head, which can cause valves to stick in their guides, with resulting rough running and loss of power. If that happens the time has come for a top overhaul. This entails removing and reconditioning the cylinder-head assembly, which is well within the capabilities of the average DIY man. One of the problems with engine repairs is that special tools are often required to dismantle and reassemble components, but all that is needed for the top overhaul job, apart from your basic tool-kit, is a valve-spring compressor. This assumes we are dealing with an overhead valve or overhead camshaft engine.

Attention to the pistons, rings, liners, bearings etc. is classified as a major overhaul and will usually entail the removal of the engine from the boat – easier if yours is a small single- or twin-cylinder model rather than a large and heavy one. Boat-builders do not always make allowance for removal of the 'machinery', so that quite a lot of work has to be undertaken to get the engine out and back in again. Some boats are literally built round the engine, necessitating removal of the wheel-house in some cases in order to lift out the engine.

Cylinder-head Assembly

The overhead valve types of cylinder-head for petrol and diesel engines are very similar in design, and the overhaul of both types follows the same procedure. (Figure 5.1)

First drain the water from the cooling system; then, after removal of the rocker box cover, the rocker shaft assembly can be lifted off when the nuts or screws securing the brackets have been removed. This gives access to the cylinder-head holding-down screws, which can be undone, allowing the cylinder-head to be removed from its securing studs. The exhaust manifold and other connections must be removed to allow the head to be lifted clear. On small engines where the weight of the head assembly is less than 50 lb. or so, it can be more convenient to lift it off with the manifold etc. attached. The gasket between the cylinder-head and the top of the block may be sticking to the joint faces, necessitating a few judicious taps with the hammer and 'old' screwdriver to prise the head clear.

Scrape off pieces of the gasket and jointing compound adhering to all the exposed joint faces of the head and the top of the block – a putty knife or old wood chisel is useful for this job. Finish off with a

Figure 5.1 *Cylinder-head assembly*

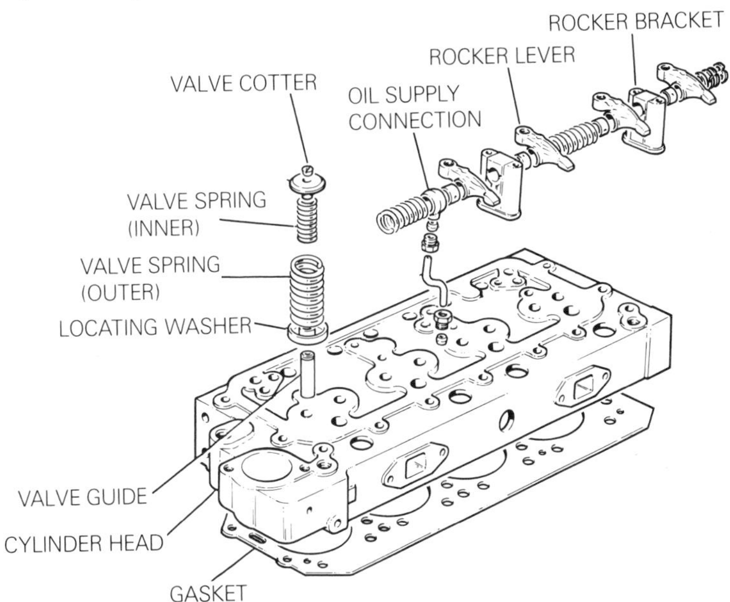

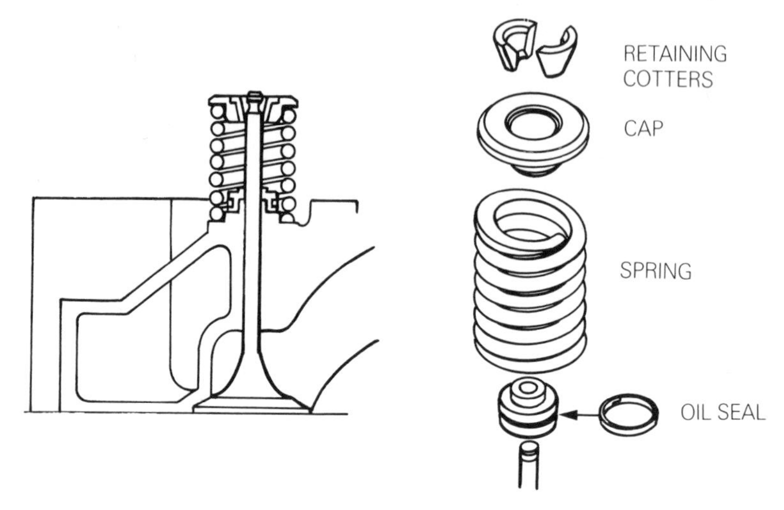

Figure 5.2 *Valve assembly*

rag soaked in petrol or solvent. Stuff rag into the cylinders and other openings to exclude dirt. Cover the top of the engine with canvas or polythene while you are dealing with the cylinder-head.

Remove cover plates from the side or end of the head, as well as the manifold and other removable fittings. Remove the valves by compressing the springs and the retaining cap so that the cotters at the end of the valves are exposed and can be removed. There are different designs of valve-spring retaining cotter – a typical split cotter is shown in Figure 5.2. Various tools are available for compressing the valve-springs – an adjustable device which fits around the cylinder-head can be purchased from a motor accessory dealer, or you can possibly use a woodworkers' 'G' clamp with a slotted adaptor for the valve end.

Don't forget to mark the numbers of the valves, starting from the front of the engine, so that you put them back in the right order – they are not all exactly the same size and shape after hammering the cylinder-head a few million times! With the springs removed, check the clearance between the valves and their guides. If this seems excessive and you see a lot of oily carbon around the valves indicative of lubricating-oil leakage, you need to fit new valve guides. Some engines have rubber oil-seals in the guides; these should be removed when refitting the valves.

Having stripped down the cylinder-head – it is not necessary to remove all the studs, incidentally – clean the carbon out of the ports, combustion chambers, etc. A blunt scraper is a useful tool for this job. Blow the ports out with compressed air – a cycle pump can be used. Wash out finally with petrol or paraffin. Corrosion in the water passages should be cleaned out with wires, and blown out with compressed air if possible.

Next examine the condition of each valve-head where it contacts the seat. If it is badly pitted or worn it will need to be refaced in the workshop or replaced by a new part. The valve seat in the cylinder-head will have to be recut if this is badly worn or burnt – you can hire a tool to do the job, though it is preferable to have this done by your dealer as you can easily scrape the head in the process. He will also check whether the springs need replacing and grind in the valves before reassembly.

Grinding-in the valves is a simple operation with the use of carborundum paste to lap the mating surfaces. The paste is smeared lightly around the valve and a suction tool is used between the palms of the hands to make the valve rotate backwards and forwards. Use medium-grade paste first and then finish with fine grade until a continuous dull grey line is produced around the seat. Clean off all remaining paste with petrol or paraffin and blow out the ports. (Figure 5.3)

Figure 5.3 *Grinding-in valves. Overhead-valve engine.*

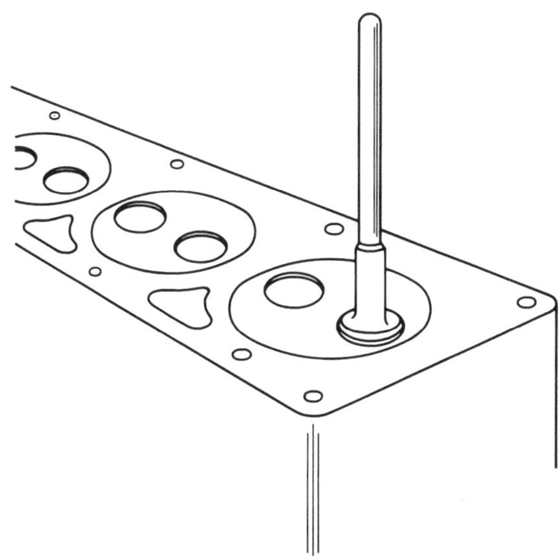

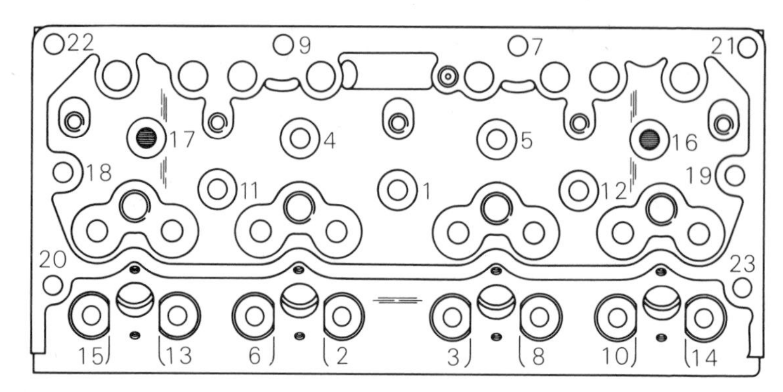

Figure 5.4 *Cylinder-head assembly. Nut tightening sequence Perkins 4.108 engine.*

You should now be ready to reassemble the valves, springs, seals and cotters, using the tool to compress the springs and allow you to insert the cotters.

Refit the cylinder-head to the engine with a new gasket, which may need to be coated with jointing compound or grease (ask your dealer if you don't have a manual). Make sure the surfaces of the block and head are clean and don't be tempted to remove the ring of carbon at the top of each cylinder. Insert the push rods in the slots through the head, so that when you refit the rocker shaft assembly you can locate the ball in the rocker lever screw into the cup of the tappet.

Refit the manifolds and carburettor, air-cleaner etc., using new gaskets.

The cylinder-head screws should be tightened in accordance with the manufacturer's diagram if you have it. If not, start with a screw to one side of the head in the centre and then work your way round, as shown in Figure 5.4. The tension of the nuts is important to assure a good gasket seal; the dealer will use a torque wrench, which gives when the desired torque is reached so that you cannot overtighten. Failing this, use a ring-spanner; work your way carefully around the tightening sequence, finally pulling up with fairly heavy pressure, bearing in mind that the length of your ring-spanner will be less than the arm of the dealer's torque wrench, so that he will exert less pressure to achieve the same result.

The actual stud torques depend on the size of the thread, the material used and the type of gasket, so that we cannot give hard and fast rules for any engine.

Timing Chain

A noisy timing chain can generally be replaced while the engine is installed in the boat, provided that there is sufficient room to work at the front end. Most petrol and some small diesel engines have a timing cover secured by setscrews to the front face of the cylinder-block. The belt-driven fresh-water pump is usually mounted immediately above the timing cover and must be dismantled, together with its connecting pipework and the crankshaft pulley, before the timing cover can be removed. Direct sea-water cooled engines, such as the Watermota, and some heat-exchanger cooled models have a sea-water pump mounted on the timing cover, so that this also has to be removed. The crankshaft pulley is usually retained by a single clamping screw; after unscrewing this you will need a puller to remove the pulley. The way should then be clear to lift off the cover with its joint, and also the oil slinger. (Figure 5.5)

Clean off any jointing compound or remains of the joint adhering to the cover or cylinder-block face. Most engines have a chain tensioner, which may not be doing its job and may thus be causing the chain to run noisily. Remove the tensioner and check the

Figure 5.5 *Timing cover. Note position of oil slinger and any spacers before dismantling.*

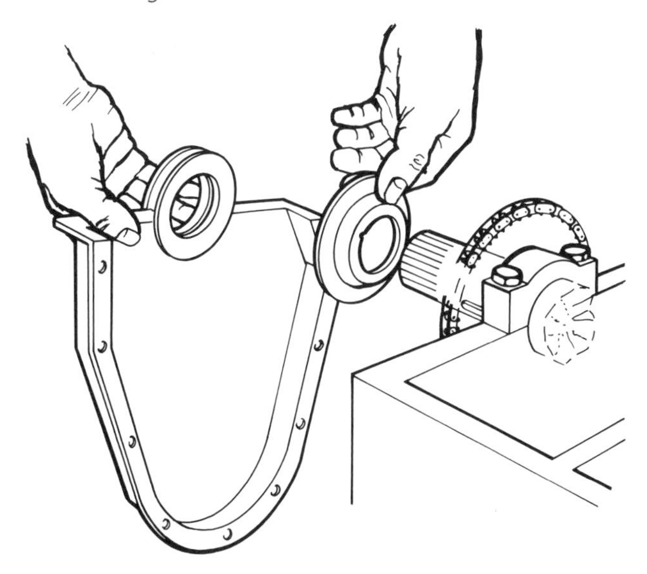

mechanism for wear. You will probably have to remove the two sprockets together with the chain, but before doing so rotate the engine (temporarily refitting the crankshaft pulley securing-screw) until the timing marks are aligned with each other. Remove the crankshaft sprocket securing-screw or screws, then lever off the sprockets with tyre-levers or something similar. Do not disturb the camshaft or crank, so that the chain can be refitted without disturbing the valve timing. Check the condition of the chain for worn or damaged rollers and slackness. Also check the sprocket teeth for wear or damage. Replace any worn components and refit. (Figure 5.6)

Examine the oil-seal fitted in the timing cover. It is good policy to replace it if the engine is fairly old, even if it has not yet started to leak. First check the surface of the crank pulley where it runs in the seal and replace if it is scored, as the seal will not function properly unless the surface is perfectly smooth. Smear the inner and outer surfaces of the new seal with grease, and press it into position as far as possible with your thumbs, remembering that the lip should face

Figure 5.6 *Timing chain. Note relative position of timing marks before removing sprockets.*

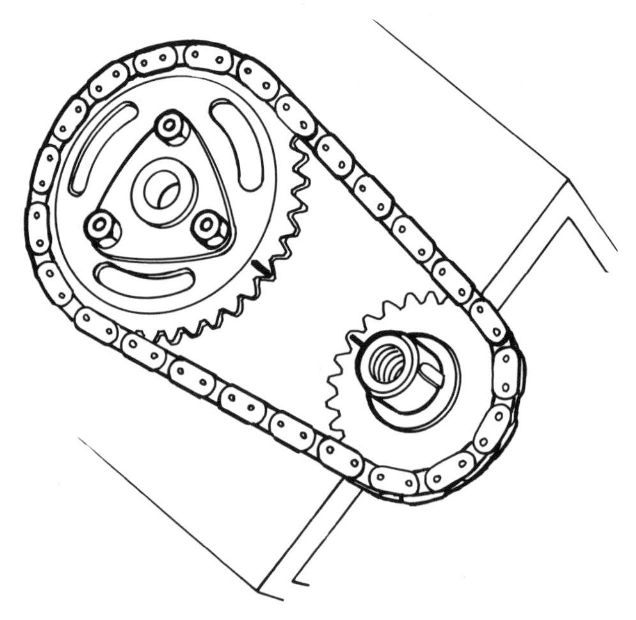

inwards. Keeping the seal perfectly square, tap it home with a mallet and piece of wood to avoid damage.

Refit the cover carefully, using a new joint and compound if required. Centre the cover on the crankshaft pulley before tightening the screws, unless the cover is dowelled in position.

Water-Pumps

The belt-driven fresh-water pump may develop a leak from the seal. Excessive sideways movement of the shaft and noise will indicate worn bearings. Most of the standard automotive type pumps cannot easily be serviced without a press or special pulley drawer, and so a new or service replacement pump is recommended.

The 'Jabsco' type of sea-water pump, however, is usually an easier proposition for DIY servicing. Impeller replacement has already been dealt with on page 63 and is a straightforward operation. Like the fresh-water pump, a new seal or shaft bearings may be required. The pump should first be removed from the engine, of course. In the model illustrated in Figure 5.7 the seal can be removed from the body after the shaft has been withdrawn with the aid of a thin screwdriver or two pieces of thin but strong wire. This seal bears on a ceramic counterface which may be worn, especially if gritty water has been pumped through. The counterface should be replaced as well as the seal. When refitting, coat the faces with grease, and observe absolute cleanliness.

Some sea-water pumps have a lip seal which bears on the surface of the shaft. Replacement is straightforward – pull out the old seal, make sure the recess is clean, then push in the new part with the lip facing the impeller. Coat the shaft lightly with grease to assist this operation.

Ball-bearings are often secured by a circlip in a groove and must be removed before pressing or knocking out the bearing assembly. Replacement bearings can usually be pressed into place in the jaws of a large vice, using a piece of tube as a spacer. Bearings should be packed with a good-quality grease unless they are of the sealed, pre-packed type.

The end-plate of the pump is sometimes worn by the passage of gritty water to the extent that the self-priming characteristics of the pump are impaired. Fit a new plate and paper gasket, or reverse the plate so that the worn part is on the outside.

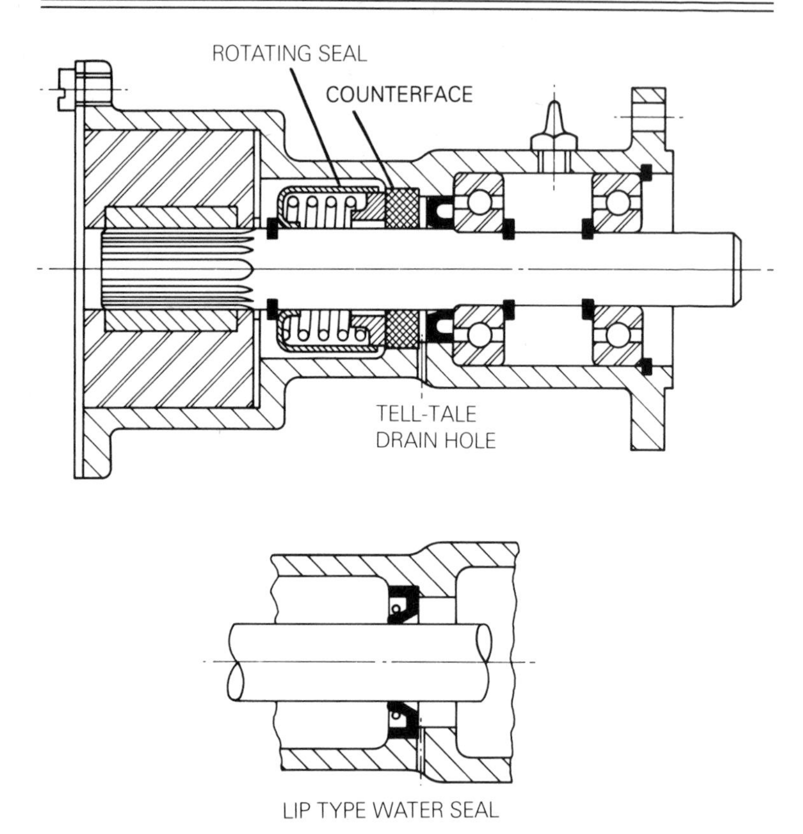

Figure 5.7 *Sea-water pump assembly*

Thermostat

If the cooling system is suspect because the engine working temperature is either too high or low, the thermostat is the likely culprit and can easily be checked. Remove this component from its housing and examine it. If it is not badly corroded, check by suspending it on a length of wire in a container such as an old saucepan filled with water. Heat gradually while observing the valve, which should start to open gradually at about 80–85°C (176–185°F) and be fully open a few degrees below boiling-point. The actual setting varies with the engine requirements and the pressure cap fitted to the heat-exchanger. If the thermostat is working satisfactorily, clean up and refit, otherwise replace with a new part. (Figure 5.8)

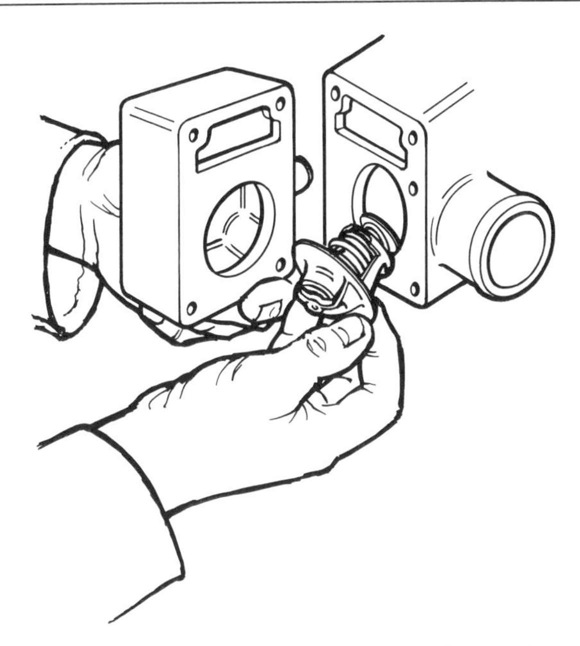

Figure 5.8 *Thermostat. Removing the fresh water bypass thermostat to check condition and functioning.*

Heat-Exchanger and Oil-Coolers

Overheating problems can be caused by blocked tubes in the heat-exchanger or lubricating-oil cooler. The gearbox oil-cooler or air intercooler of a turbo-charged engine can also be affected. The remedy is to remove and clear the blockage as explained on page 61.

Fuel-injection Equipment

Engine manufacturers agree that overhaul of the fuel-injection pump and atomizers can only be carried out satisfactorily by the manufacturer of the equipment or his agents, who will have the necessary test machines and expertise to carry out the highly-skilled operations involved.

Before removing the fuel-injection pump note the location of timing marks on the mounting flange. When the pump was originally fitted and timed to the engine a line will have been scribed across the mounting flange to correspond with a line on the engine block. A

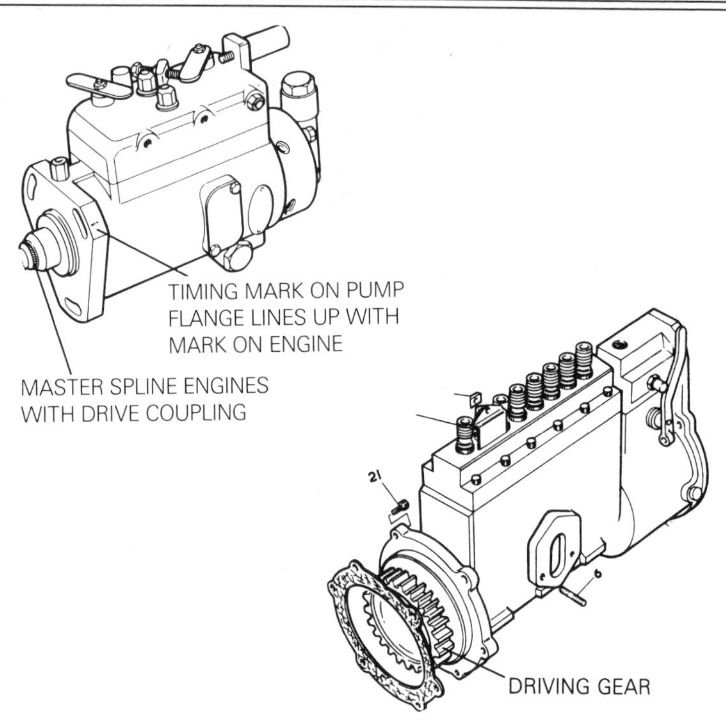

Figure 5.9 *Fuel pump timing*

master spline on the driving shaft will ensure that the mating coupling can only be fitted in one position. Other systems are also used to ensure that pumps can be timed correctly. For example, a pump fitted with a driving gear will have a mark on the teeth corresponding with the mark on the meshing timing gear. (Figure 5.9)

The fuel lift pump illustrated in Figure 2.1 on page 54 can be fitted with a new diaphragm without difficulty by removing the upper part of the pump after releasing the screws clamping the top section in position. Usually the diaphragm and its pull rod are serviced as a complete assembly. If the condition of the valves and the rocker, pin, lever etc. is poor, a new pump will be necessary.

Ignition Equipment

The plugs, leads, coil, distributor cap, points, condenser etc. should be subject to regular check and replacement as necessary, so that the only part of the ignition equipment to be mentioned here is the mechanical drive to the distributor. In time, wear may necessitate replacement of the advance and retard flyweight mechanism, or possibly the drive gears at the bottom of the shaft, one of which may be part of the camshaft. Faulty operation of the advance device, or excessive gear wear causing the ignition timing to fluctuate, will indicate a need for attention to these components.

Starter

Most starter motors – particularly the simpler Bendix type – are a reasonably easy proposition to strip down and overhaul on the garage bench. Attention to the bearings, brushes and sliding pinion with a general clean-up is straightforward, but if failure of the armature or field coils is suspected, attention by the engine-dealer or an electrical specialist is obviously required.

Removing the band clamped around the barrel by slackening the pinch-screw gives access to the brushes on some models. Worn brushes should be replaced, if they are stuck in the holders or move sluggishly; polish the sides with a file and replace in their original positions. Weak brush springs also need replacing.

Unscrewing and removing the screws holding the end-covers in position enables the armature shaft to be removed. The commutator on which the brushes bear should be cleaned with a cloth dipped in petrol.

If there is excessive clearance in the bearing bushes these need to be pressed out and replaced by new ones. A press is recommended for this operation although it may be possible to do in the vice with suitable pieces of tube as spacers. New bearings should be soaked in engine oil for a day after they have been pressed in; it is a good idea to do the same thing even if the bearings are not replaced.

The sliding pinion can be removed by extracting the split pin which locks the shaft nut. Unscrewing the nut allows the main spring and collar to be removed. Some starters have a circlip instead of nut and split pin; the spring must be compressed in order to remove the circlip.

The pinion and other parts should be washed in paraffin – any worn or damaged parts must be replaced before reassembly. (Figure 5.10)

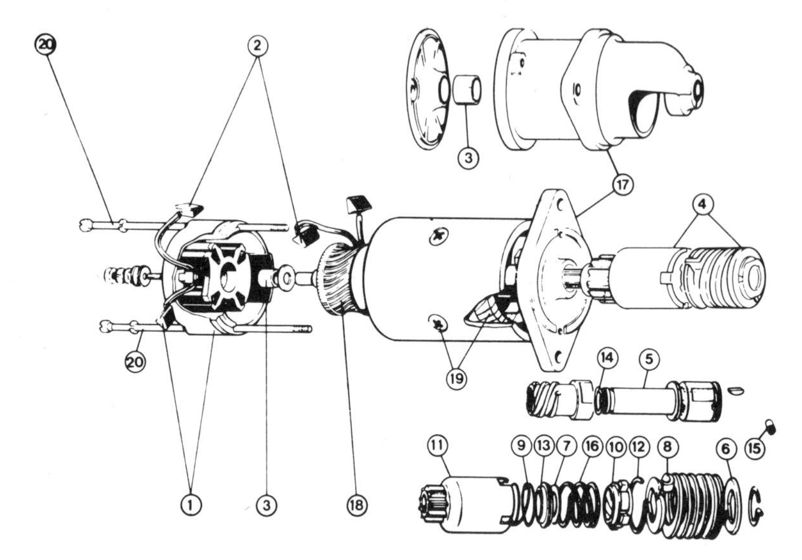

1 Bracket kit, commutator end, comprising – bracket assembly, brush and connector assembly
2 Brush kit, comprising – brush and connector, brushes (2)
3 Bush kit, comprising – commutator and bush, drive end bush and intermediate bush
4 Drive
5 Sleeve, head
6 Plate, spring anchor
7 Washer, thrust, drive head
8 Spring, main
9 Spring, cushioning
10 Sleeve, screwed, with control nut
11 Pinion and barrel
12 Ring, barrel retaining washer, thrust control nut
13 Collar, location
14 Ring, collar retaining
15 Pin, pilot drive to shaft ring, drive and sleeve retaining
16 Spring, anti-drift
17 Bracket kit, drive end, comprising – bracket, bush, dowel
18 Armature
19 Field coil kit, comprising – field coils, yoke insulator, pole screws (*2*) bracket, kit, intermediate, comprising – bracket, bearing bush
20 Through bolt kit, comprising – through bolts (*2*) lock washers (*2*)

Figure 5.10 *Starter motor*

Major Overhaul

Excessive lubricating-oil consumption is indicated by oily traces in the wake of the boat and means that the engine needs new piston rings – possibly new cylinder liners and pistons also. When your engine reaches this stage or suffers mechanical problems requiring its removal from the boat, you must decide whether to tackle the overhaul job yourself, entrust it to your local garage, get the nearest engine-dealer for your model to do it, or even buy a new engine. For the DIY man there is another alternative which is to be recommended, and that is to purchase a factory-reconditioned 'short' engine. This would be supplied in exchange for your old engine, which you do not usually have to hand over until the job is completed. You can then fit the cooling system, manifolds, carburettor, ignition or diesel fuel-injection components from your old engine. You will, of course, need to clean up and check over the items to be replaced and obtain any necessary joints beforehand.

If you feel confident enough to overhaul the engine yourself, first obtain a workshop manual for your engine. If this is a 'marinized' version of a car or truck engine, you will have to get the original manufacturer's manual. This will show you what tools are necessary, as well as explaining the various operations involved. You may be satisfied with a rebore and new pistons and rings, and a new set of main and big end bearings, if the crankshaft and camshaft etc. are not too badly worn. However, this never seems to give the same satisfaction as a completely reconditioned engine, when every component will have been checked against specification and replaced where necessary. To do the job properly you need to use a micrometer to check the crankshaft pins and journals, for example; but visual examination will show you whether new timing gears or sprocket and chain, flywheel gear ring, valve gear, and so on, need to be replaced.

A major overhaul gives you a chance to thoroughly clean the inside and outside of all components, and you may be amazed by the amount of swarf, sludge and dirt built up in the oil-ways and various internal nooks and crannies.

6 Fuels and lubricants

Diesel Fuel

Earlier chapters stressed the need to keep diesel fuel clean and free from impurities. Particular emphasis was laid on removing the water content before this could cause damage to the fuel injection equipment. This water may have contaminated the fuel during storage or it may have been due to condensation in the engine fuel-tank – particularly if stored for several months. Of course water can seep into your tank via the filler or a breather pipe, especially with flush deck fittings. If the filler is suspect this should be added to your list of end-of-season improvements to the installation. Some motor sailers when heeled over so that the side decks are awash have been known to leak sea water through the filler cap.

Getting the fuel on board and into the tank – whether you use a pump at the end of the jetty or have to transport cans via the dinghy – doesn't usually cause water problems unless you do it in a rainstorm. However, if you find the water trap needs frequent attention, check that your usual supplier has good storage facilities.

We seldom worry about getting suitable petrol for the car, but once a boat is away from the mooring and in an unfamiliar area, things can be more difficult. In the U.K., derv (Diesel Engined Road Vehicle) fuel carries more tax than the gas oil we usually purchase for boats, but it may have to be used in emergency if there is a convenient garage nearby which sells it. In an emergency paraffin (kerosene) can be used although not all engine manufacturers condone this practice, mainly because paraffin has no lubricating properties and so the close-clearance components in the fuel injection pump could seize up. The remedy is to mix a little engine oil – about 1 per cent – with the paraffin, or to use about half a tankful only to top up the gas oil already in the tank.

You must avoid the heavy grades of diesel fuel used for the large, slow-running ships' diesel engines. Some types of domestic central heating oil are suitable, however, and the 'Fuel Domestique' available in France, for example, is accepted by most engine-manufacturers. If in doubt consult your engine handbook or a local agent when you 'go foreign'.

Specifications

One thing you will have noticed is that there is no 'star' grade or octane number for diesel fuel. In fact, however, there is a minimum *cetane* number which affects the combustion properties; the specifications applying to gas oil and derv are just as stringent and detailed as those for petrol.

Fuel oil specifications cover physical properties – i.e. specific gravity, cetane value and calorific value (amount of potential energy) – and the level of impurities permitted, e.g. sulphur percentage.

Most manufacturers of 'high-speed' diesel engines recommend the use of fuels to the following specifications in the countries quoted:

United Kingdom BS2869 : 1967 Class A1 and A2
United States A.S.T.M./D975 – 66T – Nos. 1–D and 2–D
Germany DIN – 51601 (1967)
France J.O. 14/9/57 Gas Oil
Italy Cuna – Gas Oil NC–630–01 (1957)
Sweden SIS 155432 (1969)

Many other countries have, or are establishing, similar specification standards.

Cold Weather

Few pleasure-craft engines are operated when the temperature drops below freezing-point, but many fishermen and other commercial boatmen have to cope with much lower temperatures. Under these conditions the fuel system of a diesel engine can become blocked by the wax which is precipitated unless the tanks, piping and filters can be kept warm – not usually a practical proposition.

In the UK and most other countries, fuel for diesel-engined road vehicles is sold in summer and winter grades. The winter grade is produced by changes in the refining technique, and by the use of additives to reduce the wax precipitation temperature. It is possible to improve the low temperature characteristics of summer fuel by simply adding paraffin (kerosene) – up to 15 per cent is recommended by the oil companies for this purpose. Petrol may also be used up to about 30 per cent.

Alternative Fuels

There is much concern about dwindling supplies of the crude oil from which diesel fuel and petrol are distilled. From time to time various estimates are made of how much 'fossil fuel' supplies exist and what will happen when they are exhausted. Alternative fuels made from many substances are being tried out; most cost considerably more than what we are currently using, but as current costs increase the alternatives become more viable. Organic fuels made from sugar-cane and other plants are being manufactured in countries such as Brazil where they do not have adequate supplies of crude oil, but where they are able to grow agricultural products in vast quantities.

The engines in our boats today run happily on the fuels currently available, with some paraffin or heating fuels mixed in occasionally when we cannot get the specified gas oil. But some changes in the engines will be necessary to make them more tolerant of fuels outside the present specifications – fuels that allow the oil refineries to get more gallons from a barrel of 'crude'. Using the vegetable-derived fuels, except when mixed in small quantities with a conventional fuel, will also necessitate engine modifications. Engine manufacturers are experimenting with new combustion systems and fuel injection equipment capable of using such fuels of the future.

Perkins have developed a new combustion system called 'Squish Lip' which, as well as reducing combustion noise and stresses in the working parts of the engine, can use fuel which is 'degraded' below the present standards.

However, although temporary fuel shortages are likely to continue, supplies seem to be assured well beyond the life of today's engines. One thing we can be sure of is that fuel prices will continue to rise – and so economical operation of the engine must be assured by maintaining it in good condition.

Lubricating-oils

We can dismiss this subject very briefly by saying: 'Use the makes and grades of lubricating-oil specified in your engine operator's manual, and change the oil and filters at the specified periods'.

A little background knowledge may come in handy, however, especially when you can't get hold of the specified oil. Some of the questions which occur are:

'Why do petrol and diesel engines require different grades of oil?'
'Is it detrimental to the engine if I top up with a different grade or brand?'
'How do multi-grade oils acquire this characteristic?'
'Is it safe to exceed the oil change period occasionally?'
'Why do oil change periods vary from engine to engine?'

The following notes will explain these and other questions and may help to give you more confidence in your ability to avoid problems associated with your lubricating-oil.

Oil Grades

You will have noticed that while marine petrol engines – at least the four-stroke type – generally specify the familiar brands and grades of oil sold for use in cars, oil for diesel engines has less well-known designations although produced by the same oil companies. The reason for this is that diesel engines need a 'heavy duty' oil because the lubrication of these engines is a more difficult proposition. The pressure squeezing the oil film in the main and big end bearings will be higher; temperatures may be higher and hours of continuous operation longer.

Also, the impurities which the oil has to collect in its passage through the engine will be more difficult to deal with; the oil should hold solid particles in suspension. A good quality oil will also protect the bearings and other surfaces from corrosion (especially when the engine is not running); it will inhibit rust formation, will not foam excessively and will not decompose or thicken prematurely. It should also retain its 'oiliness' and reduce friction as well as wear, scoring and seizure.

A modern oil, whether for petrol or diesel engines, which displays these properties starts life as a good quality refined petroleum crude oil into which an additive 'package' has been mixed.

The ability of oils to carry out the functions mentioned can only be ascertained by practical testing in an engine. A number of engines are regularly used for oil testing by oil companies, specialist firms and government agencies, and a relatively small number of engine types are employed for this purpose.

The many brands of oil on the market – at least the ones which engine-builders approve for use with their products – comply with the testing authorities' standards. For example, the United States Military specifications MIL–L–46152 or MIL–L–2104C are acceptable to most manufacturers of high-speed diesel engines.

Oil Change Periods

The testing carried out by engine-builders under actual working conditions – in pleasure-craft and work-boats of various types – determines the specified oil change period. The way the engine is used – for example, continuous 'flat-out' performance for less than 100 hours per year; regular steady use at half-throttle or less for, say, fifty hours a week; or fifteen minutes twice a day – makes a big difference to the hours of duty between oil changes. However, the operators' handbook cannot go into this level of detail and to simplify matters will generally quote, say, 150 hours between changes, whatever the type of craft or duty cycle.

It may be, therefore, that you are changing oil more frequently than necessary. For pleasure-craft once a season is the minimum, although the actual hours in an auxiliary yacht, for example, may be substantially less than the recommended figure in the book.

Commercial operators of large engines running significant hours may have the engine's oil condition analysed at regular intervals by the oil company's laboratory to see whether the particular work cycle entails an oil change as specified, or if the 'service hours' can be prolonged. In certain cases the normal period needs to be reduced, but this is rare.

Multi-grades

The viscosity, i.e. thickness of the oil at a certain temperature, is specified – usually to SAE standards so that we have a thinner oil in winter (SAE 10, say) and a thicker one in summer (SAE 30, for example).

If we leave the thin 'winter' oil in the engine during the summer it will not lubricate properly and the oil pressure may be low. Also, if the thick summer oil is used in winter we may not get the engine to start because of the increased oil 'drag'. Multi-grade oils which are intended not to become too thin when hot or too thick when cold are now commonplace.

If we are forced to mix oils we may be combining additives which react with each other, although both oils may comply with the same standard. However, if it comes to a choice between running short of oil and topping up with 'Brand X', do so, but make sure you have adequate supplies of a single brand on board after the next oil change.

Synthetic Oils

Internal combustion engine technology never stands still, and the manufacturers of lubrication products strive to produce better oils by developing new additives to keep engines in better condition for longer operating periods.

By manufacturing the oil from 'synthetic' materials rather than starting from a 'natural' crude oil, it has been found that a more stable oil which will run for longer service periods is possible. At the present time synthetic oils are much more expensive than the 'natural' type, so they are really only of interest for very high performance engines. But as crude oil products become more expensive and the manufacture of synthetic types is developed we may find them competing with the 'natural' oils in a few years' time.

Two-stroke engines

Petrol fuel for these engines, which do not have a separate lubrication system, can be purchased ready-mixed; or you can mix it yourself. It is important not to use too much oil – if you do, the plugs will oil up – and also to shake up the fuel tank if it has been standing unused for some time. You may find that slightly less oil than the specified proportion is beneficial when the engine is fully run in, but check with your dealer and don't 'overdo it'.

There is a danger of fouling the plugs if you mix in oil with unsuitable additives. Most outboard manufacturers recommend a straight monograde oil sold expressly for two-stroke use.

Oil-cooling

Engine builders set a maximum temperature for the lubricating-oil at the designated measuring point, which might be the sump, the filter or in the pressure rail of the cylinder-block. From tests the maximum temperature at the critical parts of the engine to which the oil circulates will have been measured – these may be the main or big end bearings. By keeping the oil within the specified limit at the measuring point, the oil pressure and thus the oil film in the bearings will be protected.

In many cases it will be necessary to cool the oil by means of the circulating cooling water. A cooler as described in Chapter One is used for this purpose.

Transmission Oil

Hydraulic gearboxes use oil pressure to operate the clutches, and they lubricate the gears and bearings by means of the same oil. An oil-cooler is used in some cases where the temperature would otherwise be excessive.

The specified type of oil used is frequently the same as the engine oil, and this is obviously beneficial: it prevents the wrong type from being used in the engine or gearbox, and makes only one container necessary for topping up supplies.

A non-foaming oil such as an automatic transmission fluid is preferred by some gearbox manufacturers as the breather may be subject to oil leakage.

Because there is no carbon or other products of combustion present in gearbox oil, the service periods are usually much greater than for the engine oil. After the initial period when the gears and clutch-plate have 'run in', the oil is changed and afterwards will usually remain fresh and clean-looking through the season. This is very different to the appearance of engine oil, which usually goes black in a few hours as it absorbs soot – i.e. carbon ash – from the combustion area.

Filtration

Filters for both the fuel and lubricating oils are described in Chapter One. The 'final' filters in both systems usually embody a paper element and engine builders insist on 'genuine' parts being used. This is not just sales patter. The major engine builders test filter elements regularly to ensure that they trap particles of impurities down to the specified levels and do not collapse under the operating pressure allowing impurities to damage the fuel injection pump or the engine bearings. It is certainly 'false economy' to use inferior filter elements.

Bypass Lub-oil Filters

The lubricating oil filter is almost always a full-flow type i.e. the entire delivery from the oil pump passes through the filter en route to the bearings and other parts of the lubrication system.

Although the standard filter is considered adequate by the engine builder for most applications, it is possible to extend the life of the filter element and the time between oil changes by using a bypass filter. This device takes a 'bleed' of oil from the main circuit and passes it through a fine filter of large capacity. Filters of this type are sometimes installed on large diesel engines powering fishing vessels where the solid contaminants and corrosive acids are removed from the oil so that the oil life is extended almost indefinitely.

7 Safety precautions

The use of non-slip surfaces, guard rails and safety harnesses is common practice for the deep-sea sailor. Safety measures below decks are just as important, however.

Fire Prevention

Fighting a fire at sea is the most important trouble-shooting you could be called upon to handle and so it makes sense to be well prepared for this contingency.

Fires in the engine compartment are frequently caused by a fractured pipe discharging fuel on to a hot surface. Some Classification Societies have rules which call for high-pressure fuel-pipes to be double-skinned or enclosed by a sleeve, so that if the pipe fractures fuel is directed down the sleeve, where it runs into the bilges or (preferably) a tray under the engine. Fires caused by petrol engines that spit back are guarded against by the use of flame traps on the carburettor. These are mandatory for many commercial craft and to meet the standards of Fire Offices. Poor lagging on hot exhaust pipes can allow woodwork in the vicinity to catch fire. Failure of the sea-water pump cuts off the supply of coolant to the exhaust system, and where rubber hose is used this will catch fire. Less commonly, electrical fires may result from a short circuit in the wiring system or control board.

The first requirement, of course, is to have adequate fire-fighting equipment on board. There are many brands and types of fire-extinguisher which can be used, although not all are suitable for dealing with the types of fire which can break out in the engine compartment or elsewhere on the boat. A minimum of two hand-operated extinguishers should be carried, positioned at the fore and aft ends of the boat. The size of these extinguishers should be appropriate to the boat, and the UK Fire Offices Committee recommends the size of extinguishers, which would vary, for example, according to whether there is a galley with gas cooker installed.

For the engine compartment, where the US coastguard records show that 95 per cent of ship-board fires occur, the best plan is to have an extinguisher that is triggered automatically when a fire breaks out, or alternatively one that can be operated from outside the compartment. You do not want to have to lift the hatch covers before you can deal with the fire, as this would cause it to burn more fiercely because of the extra air supply you have provided.

Fire-extinguishers are filled with several different materials, each operating in a different way. Dry powder, which is usually based on bicarbonate of soda, acts by chilling the flames in the same way as water. It is suitable for quenching ship-board fires – including electrical ones – but tends to make a mess, which may not matter if you have successfully fought a major outbreak but might be considered a disadvantage when dealing with a minor one. Another snag is that, in time, vibration caused by the engine makes the powder pack down in the container, so that it is not expelled by the propelling gas when you pull the trigger. As with other types of extinguisher, you must check the dry powder type regularly.

Carbon dioxide gas is another suitable fire-extinguisher. This works by cutting off the air supply to the fire by displacing it, i.e. smothering the flames. CO_2 is suitable for built-in systems and is very effective, although it can short-circuit exposed contacts in an electrical panel by depositing ice, formed by the gas expanding as it leaves the nozzle. The cold temperature can also crack exhaust manifolds and turbo-charger casings if directed onto the hot surface, so that the location of discharge points must be carefully considered.

It is generally considered that one of the best systems for boats uses a gas with the lengthy name of Monobromotrifluoromethane. The proprietary name is Halon and this is manufactured in the USA by Dupont and in the UK by I.C.I. The gas is blended with other chemicals to improve its speed of vaporization when it emerges from the extinguisher, where it is stored under pressure as a liquid. Like CO_2, Halon is very suitable for use in automatic extinguisher systems because it works by chemical action in combining with the

Figure 7.1 *Automatic fire-extinguisher. Golden Arrow Extinguishers Ltd.*

vapours produced by the flames, turning them into compounds which do not burn. Halon will penetrate to all parts of the engine compartment or other areas to which it has access, and it does not chill the surfaces contacted.

Foam is a familiar fire-extinguishing and fire-prevention medium used by airports. It is also used for fire-fighting around docks and marinas, where it is suitable for operation with both fresh and salt water. Foam is used in commercial vessels but rarely in pleasure-craft. Various foaming agents are used to make water float on top of flammable liquids, forming a vapour seal to snuff out the flames.

Automatic extinguisher systems are triggered by temperature. In the engine compartment this is usually set between 73°C (165°F) and 100°C (212°F). The position of the actuating valve is important: it must not be located adjacent to a hot spot or the contents will be discharged unnecessarily. A very simple automatic extinguisher can be obtained, as shown in Figure 7.1, and this is quite compact and suitable for attachment to a bulkhead or other surface in the engine compartment. For larger machinery installations there would be a system of pipes discharging through several strategically placed nozzles. As an alternative to the automatically discharging system, manual operation of a fire-extinguisher placed outside the engine compartment can be arranged. The extinguisher would be connected to nozzles within the engine compartment and areas of the bilges etc., as indicated in Figure 7.2.

The larger installations with several nozzle outlets are operated electrically, pneumatically or manually, different systems being favoured by the various suppliers. The two systems shown in Figures 7.1 and 7.2 are manufactured and distributed by Golden Arrow (Extinguishers) Ltd. of Newhaven, England.

Figure 7.2 *Manual fire-extinguisher system. Golden Arrow Extinguishers Ltd.*

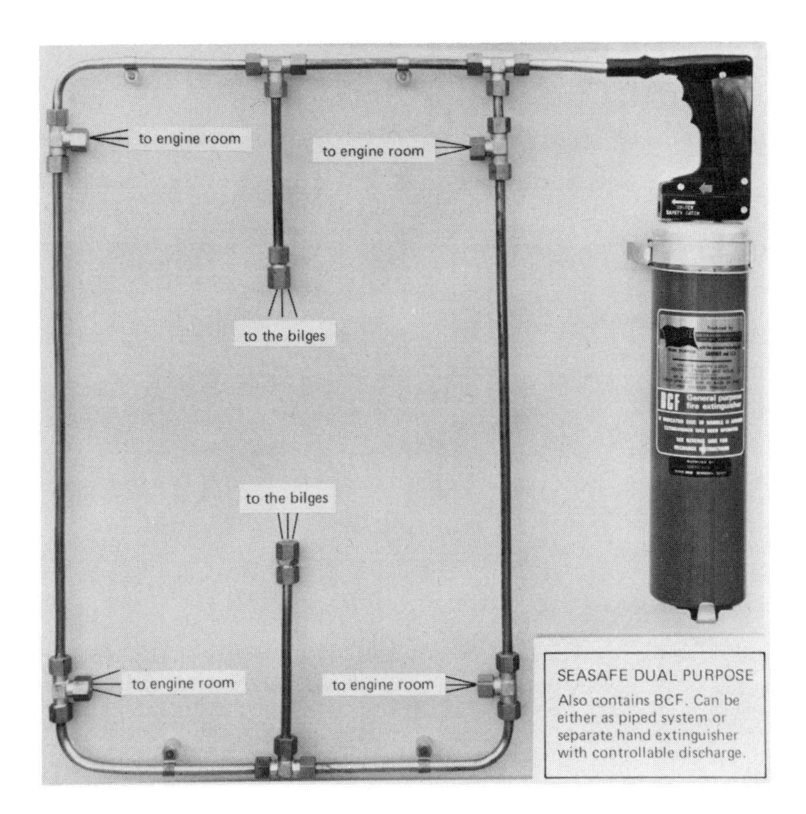

Engine Starting

Many modern petrol and diesel engines – especially larger models of 30 bhp and above – cannot be hand-started and they thus rely absolutely on the electric starting system. In the event of a breakdown at sea this may fail to restart the engine, and so the boat is rendered helpless.

For lifeboats and some commercial vessels an alternative system for starting engines is mandatory. This is often specified as a hydraulically operated cranking motor actuated by 'air over hydraulics'. An accumulator is used to store the energy provided by air or an inert gas, which is pumped up by hand in the first instance. With the engine running, pressure can be maintained in the accumulator by means of an engine-driven pump. (Figure 7.3)

Hydraulic starters may be considered an expensive luxury on a pleasure-craft, unless of course you experience conditions which knock out the electric starting system. The battery is usually the weak link in the chain. It may be damaged by violent motion of the boat – especially if not properly installed. It may fail just when you need it or not provide sufficient power to get the engine restarted, especially if you have to purge air out of the fuel system by motoring the engine over for several seconds.

The answer is to have on board a fully-charged stand-by battery,

Figure 7.3 *Hydraulic engine starting. American Bosch hydrotor system.*

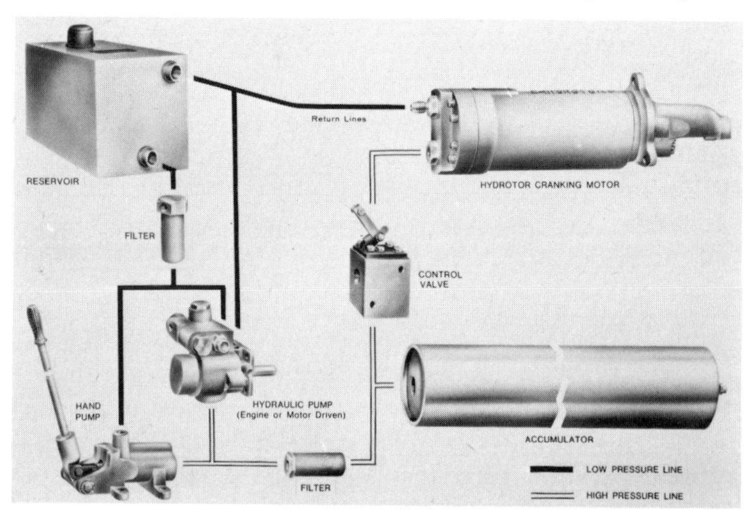

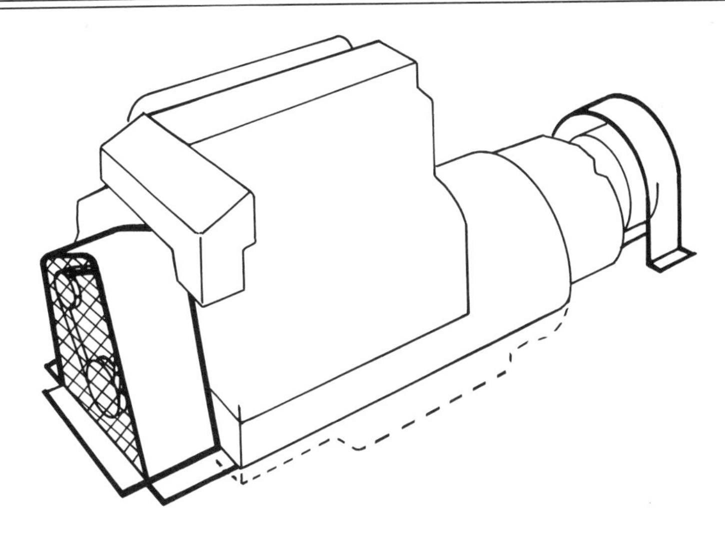

Figure 7.4 *Safety guards*

stowed carefully in place where it can be connected in parallel to the starting battery terminals by jump leads. You may be able to bring the domestic supply battery into use for an emergency start, but the chances are that this will not be fully charged, and in any case for diesel engines it will be of a type to provide adequate capacity rather than high starting performance, as described in Chapter One.

Safety Guards

Factory regulations ensure that machinery is adequately guarded, and this requirement also applies to places of work such as commercial vessels. The owner of a pleasure-boat has no such legal obligation, although negligence causing injury to a crew member could result in a law suit. It is advisable to ensure that rotating parts of the engine and the propeller-shaft, couplings etc. are fitted with guards when necessary, to avoid personal injury. It is not usually possible to provide guards as standard equipment on the engine, because items such as belt drives, power take-off shafts and couplings are seldom fitted up until the engine is installed. If the boat-builder has not made adequate provision, then this is something the owner should do at the earliest opportunity. Figure 7.4 shows typical guards for an engine installation.

Lead-Acid Batteries

The following notes are provided by the Chloride Group Ltd.

Lead-acid batteries are safe in use provided that certain simple precautions are observed. In order to ensure safety it is essential that the battery be properly operated and maintained in accordance with the manufacturer's recommendations.

Acid

The battery contains dilute sulphuric acid which is poisonous and CORROSIVE. It can cause burns on contact with skin and eyes. It is advisable to wear protective clothing and in particular to protect the eyes.

If acid is spilt on skin or clothing, wash with plenty of clean water. If acid gets into the EYES, wash well with plenty of clean water and get IMMEDIATE MEDICAL ATTENTION.

Gases

Batteries give off EXPLOSIVE gases. Keep sparks, flames and lighted cigarettes away from the battery. SWITCH OFF CIRCUIT before connecting or disconnecting the battery, as otherwise a spark can cause an explosion.

Ensure connections are tight before switching on.

Areas where batteries are kept or charged must be adequately ventilated.

Electricity

The very heavy currents which flow when battery terminals are incorrectly connected may cause molten metal to spit out. Never allow metal objects to rest or fall on the battery terminals. Before a connection is made to the battery the circuit must be checked to ensure that it is safe.

ALWAYS PROTECT THE EYES

8 The transmission system

Although most boat engines have a reversing gearbox with a shaft drive to the propeller, the advantages of alternative arrangements have led to their increasing popularity, and so notes on the maintenance of various systems are given in this chapter. Though these will serve for general guidance, the different designs used by the manufacturers may necessitate different procedures, so it is always best to refer to the supplier's handbook when this is available.

Reverse/Reduction Gearbox

Reversing gearboxes are usually supported by the engine flywheel housing and driven via a flexible coupling, as illustrated in Figure 8.1. Some engines which run at a suitable speed for the propeller do not require reduction gear, and this simplifies the arrangement, as shown in Figure 8.2, where the reduction gear is attached to the aft end of the reverse gear. In the 'drop-centre' type of gearbox, illustrated in Figure 8.3, the reduction gear is incorporated in the drive to the output shaft.

Virtually all models in use today have a lubrication system which is independent of the engine and thus keeps free of carbon and other combustion impurities. The more sophisticated hydraulic type has an oil-pump supplying pressure to operate the clutches, which engage ahead, reverse or neutral in the epicyclic gear. The second box has splash lubrication and is mechanically operated,

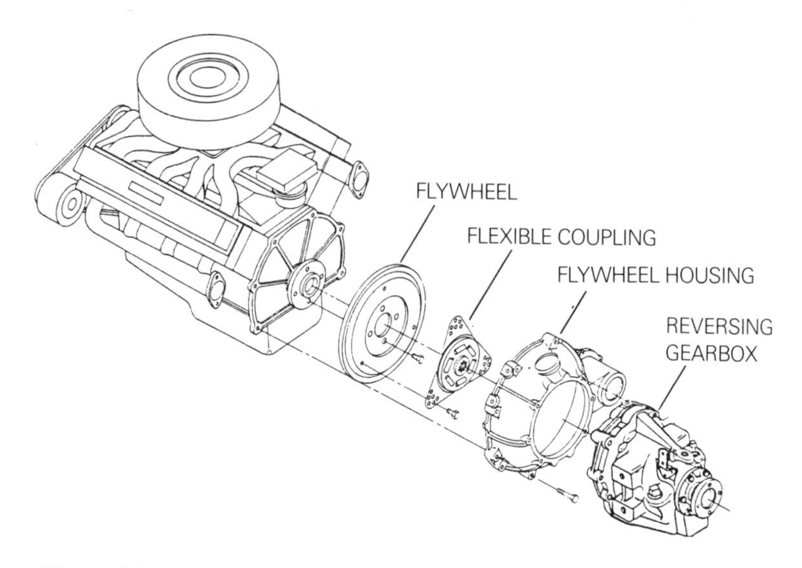

Figure 8.1 *Transmission system*

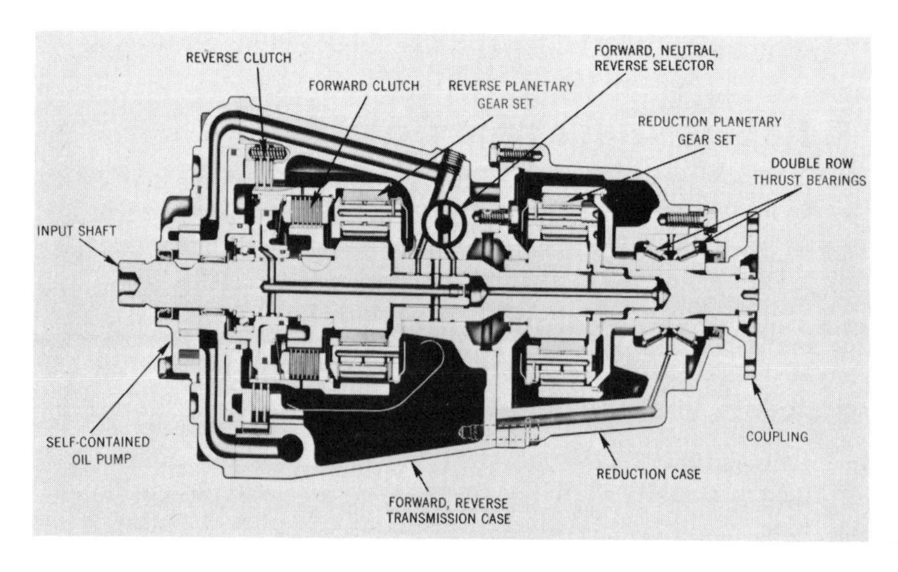

Figure 8.2 *Reverse-reduction transmission (Borg Warner)*

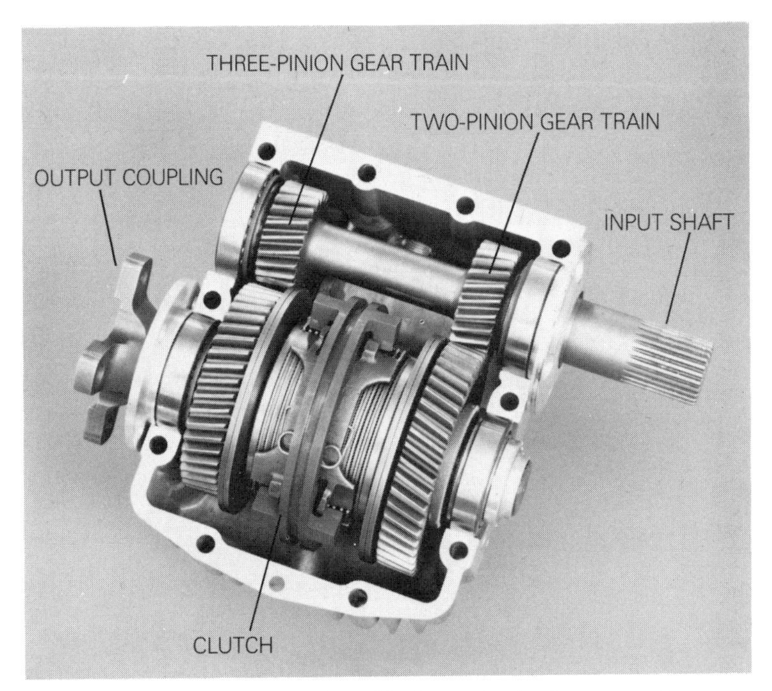

Figure 8.3 *Drop-centre gearbox. Hurth mechanically-operated type.*

either the forward pair of gears or the reverse train (which includes a third pinion) being engaged by the operating lever.

Both gearboxes have self-adjusting clutches, and the only routine maintenance required is to check the oil level and top up when necessary. Look out for excessive noise or overheating, which might indicate clutch slip, a bearing failing or other problems which should be attended to by your dealer as soon as possible.

Hydraulically operated transmissions usually have an oil-cooler external to the gearbox and are sea-water cooled. Oil connections to and from the gearbox are high-pressure flexible hoses, which must be inspected occasionally for signs of perishing and leakage of the end connections.

Borg Warner publish trouble-shooting charts which enable problems to be diagnosed so that a simple fault such as a leaking plug or incorrect control adjustment can be dealt with immediately. More drastic problems necessitating stripping the gearbox are best dealt with by the manufacturer's dealers, who have the necessary equipment as well as a stock of replacement parts.

Trouble-Shooting Chart (CR2 Model)

COMPLAINTS & SYMPTOMS	REMEDY	
	Transmission in Boat	*Transmission Removed*
INTERNAL & EXTERNAL LEAKS		
1. Oil leaks at pump		1 2 3 8
2. Oil on exterior of trans.	1 4 6 7	2 5 8
3. Oil leaks at rear seal	3*	
4. Water in transmission oil or oil in cooling water	9	
5. Oil leak from breather	9 15 19 43	
TRANSMISSION MALFUNCTIONS IN ALL RANGES		
1. No oil pressure	10	11 12
2. Low oil pressure	13 14 15 16 18	17
3. High oil temperature	9 15 19 20 21 22 30	17 42
4. Failure of reduction gear	21	23
TRANSMISSION MALFUNCTIONS IN FORWARD RANGE		
1. Low oil pressure	13 14 15 16 18	17
2. Forward clutch engages improperly	37	12 20 24 25 26 27 28
3. Forward clutch drags	37	26 27 28
4. Reduction unit failure		23
TRANSMISSION MALFUNCTIONS IN REVERSE RANGE		
1. Low oil pressure	13 14 15 16 18	17
2. Reverse clutch engages improperly	37	24 26 28 29
3. Reverse clutch drags	37	26 28 29
4. Reverse gear set failure		42
5. Reverse gear set failure		23
TRANSMISSION MALFUNCTIONS IN NEUTRAL		
1. Ouput shaft drags excessively in forward position	37	26 27 28
2. Ouput shaft drags excessively in reverse rotation	37	26 28 29 42
MISCELLANEOUS TRANSMISSION PROBLEMS		
1. Regulator valve buzz	15 16	
2. Gear noise – forward	31	32
3. Gear noise – reverse	31	32 42
4. Pump noise	15	17 32
5. Damper noise or failure		33 34 35 36
6. Shifts hard	7 16 37 39 38	
7. High oil pressure	16 30 40 41	

* If installation allows access, otherwise remove transmission.

KEY TO TROUBLE-SHOOTING CHART

Item 1 Loose bolts – tighten
Item 2 Damaged gasket – replace
Item 3 Damaged oil seal – replace
Item 4 Oil line fitting loosened – tighten
Item 5 Case leaks, porosity – replace
Item 6 Oil filter plug leaks – tighten or replace
Item 7 Damaged control valve 'O' ring – replace
Item 8 Foreign material on mating surfaces – clean
Item 9 Damaged oil cooler, water and oil mixing – replace
Item 10 No oil – find leak and fill
Item 11 Pump improperly located for engine rotation – locate correctly
Item 12 Sheared drive key – replace
Item 13 Faulty oil gauge, replace, bleed air from gauge line
Item 14 Dirty oil screen – clean or replace
Item 15 Low oil level – add oil to proper level
Item 16 Regulator valve stuck – polish with crocus cloth to remove burrs and clean
Item 17 Worn oil pump – replace
Item 18 Regulator valve spring weight low – replace
Item 19 High oil level – drain oil to proper level
Item 20 Low water level in cooling system – fill
Item 21 Dirty oil cooler – clean or replace
Item 22 Cooler too small – replace with larger cooler
Item 23 Inspect reduction unit – repair
Item 24 Worn or damaged clutch piston oil seals – replace
Item 25 Worn or damaged clutch sealing rings – replace
Item 26 Clutch improperly assembled – rebuild
Item 27 Damaged or broken Bellville springs – replace
Item 28 Worn or damaged clutch plate(s) – replace
Item 29 Damaged or broken clutch springs – replace
Item 30 Cooler lines damaged or too small – replace
Item 31 Inadequate torque on output shaft nut – tighten
Item 32 Nicks on gears – remove with stone
Item 33 Excessive runout between engine housing and crankshaft – align
Item 34 Wrong damper assembly – replace
Item 35 Damaged damper assembly – replace
Item 36 Body fit bolts not used in mounting holes – replace
Item 37 Control linkage improperly adjusted – adjust
Item 38 Control lever and poppet ball corroded – clean and lubricate
Item 39 Control linkage interference – check and adjust
Item 40 Wrong oil used in transmission – change
Item 41 Cold oil
Item 42 Planetary gear failure – replace or repair
Item 43 Damaged breather – replace

Controllable-pitch Propeller

By twisting the angle of the propeller blades the pitch – i.e. *the theoretical distance moved forward by the propeller for each revolution* – is altered.

A conventional fixed-pitch propeller is usually designed to be most efficient when the engine is run at full throttle and is comparatively inefficient at other throttle settings. By being able to vary the pitch, therefore, you can obtain greater efficiency and so better fuel economy. In practice the improved economy is more important in large commercial vessels; generally more significant for a small boat is the ability to control the speed precisely by means of the pitch control, when attending to lobster pots for example. Of course the ability to twist the blades round to the 'opposite hand' position and thus eliminate the reverse gearbox is possibly the most useful attribute of all. (Figure 8.4)

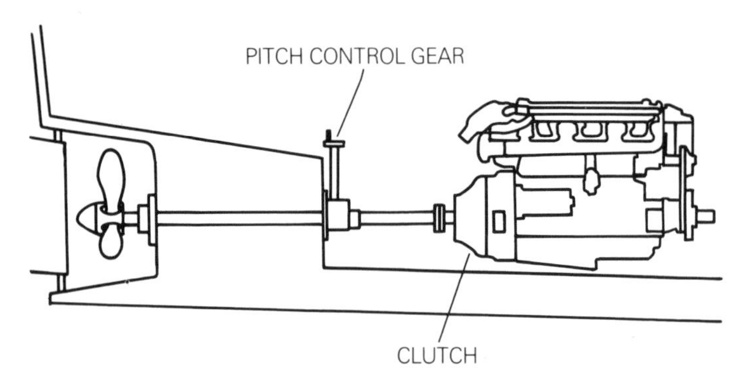

Figure 8.4 *Controllable pitch propeller*

Unless your engine speed coincides with an acceptable propeller-shaft speed you will need a reduction gear, so you cannot usually eliminate the gearbox altogether.

Designs of C.P. propellers vary – usually the propeller-shaft is tubular, with the pitch control rod sliding within to engage the mechanism in the propeller hub to twist the blades.

Servicing the C.P. propeller assembly consists mainly of regular greasing or oiling by the means provided, examination of propeller-shaft bearings and shaft gland packing, and checking and replacing as required the propeller-blade seals in the hub.

Stern-drive

This is in principle an elongated gearbox with a bevel gear drive to the propeller-shaft. As it is located outboard of the transom it is exposed to salt-water corrosion, damage from the mooring or other boats, and underwater leakage of sea water in, or oil out, depending on relative levels. There is the added complication that the steering action may also tilt (or swivel) to the 'parked' position. The gear change mechanism is usually incorporated with the upper bevel gears – sometimes in the bottom end. (Figure 8.5)

There are several manufacturers and they recommend slightly different maintenance procedures. Essentially, though, it is a matter of maintaining the correct oil level, as most are splash-lubricated, and of checking for the presence of water, which would indicate a leak at the bottom end or possibly in the rubber gaiter which seals the swivelling joints in the drive.

Because the external components are of aluminium, zinc sacrificial anodes are generally used to protect the casing etc. from

Figure 8.5 *Stern-drive BMW with 6-cylinder petrol engine (BMW)*

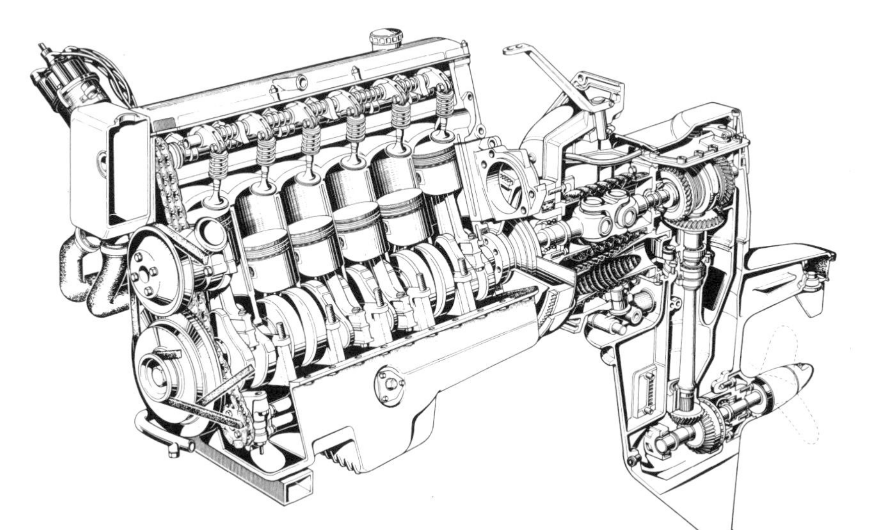

corroding away. A renewable zinc ring, located behind the propeller, is usually provided for this purpose.

At the end of the season it is recommended that the drive is removed from the boat and stored inside. (While it is being removed, make sure that the hole in the transom is sealed up as well as the back end of the engine.) This gives you an opportunity to clean the unit down and examine it for damage to fittings, control cable attachments etc. Repair any damage to the propeller – take it to a repair specialist if there is more than just a slightly bent blade. Pump water-resistant grease through all the grease-nipples provided, but don't overdo it. Clean the cooling water intake grille if this is incorporated in the leg. Drain the oil and check for any creamy consistency, giving warning of water contamination and necessitating a close examination of seals and the gaiter. Grease the engine drive shaft spline and the attachment bolts. EP90 oil is generally recommended for stern-drives.

If you need to touch up the paint on the drive casing, use the original type (obtainable from a dealer) that is compatible with the rest of the drive surface. Most drive legs can be painted with anti-fouling to prevent weed growth. An epoxy type is usually recommended by the manufacturer.

Sail-drive

This arrangement is similar in operation to the stern-drive but is less complicated because it does not steer or tilt. Maintenance is simple – generally it is not considered necessary to remove it from the boat at the end of the season, and routine attention consists of keeping an eye on the oil level and its consistency so as to check for contamination and leaks. The rubber sealing diaphragm should be examined occasionally as this is the means of keeping water out of the boat. In the design illustrated, a double diaphragm is used with a sensor located between the two sections to give warning of leakage past the lower part. A warning buzzer on the engine panel is triggered by the sensor. Any propeller damage or paint touching-up can obviously be tackled when the boat is hauled out, and the sacrificial anode can be replaced if showing signs of being eaten away. The oil is changed at the end of the season or after 150 hours' service in the majority of sail-drive models. A first oil change at about 25 hours is recommended to remove debris from the running-in of the gears. (Figure 8.6)

Figure 8.6 *Sail-drive Perkins 4.108 (Perkins)*

Hydrostatic Systems

'Hydraulic drives' are very convenient when it is desirable to locate the engine away from its normal situation in front of the propeller-shaft. A transverse position is used in the stern of hire craft, for example, with only the hydraulic motor at the inboard end of the prop-shaft.

The design of the hydraulic motor is such that by selecting various models the drive ratio of the pump to the motor shafts can be varied and a speed reduction achieved, as in a conventional reduction gearbox. (Figure 8.7)

These systems operate at high pressure and the flexible hoses specified by A.R.S. Marine Ltd. for their hydraulic drives are designed to withstand a bursting pressure of 16,000 pounds per sq. in. (1120 Kgf./cm.^{2}). Maintenance of an A.R.S. system is recommended as follows:

Change the hydraulic oil filter every 50 hours.

Check hoses for external wear every 50 hours.

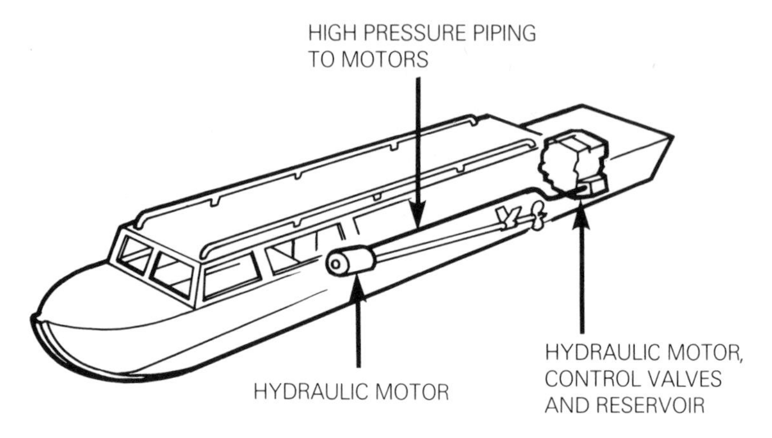

Figure 8.7 *Hydrostatic drive system*

Change the oil every 400 hours or two years.

Oil type: Castrol AWS 32 or Shell Tellas 27.

When starting up at the beginning of the season, and especially after the system has been drained completely, the engine should be turned over slowly – by hand or on the starter – to prime the hydraulic pump unit and ensure lubrication of the motor bearings. On starting, run the engine at quarter throttle for 3–4 minutes, with the hydraulic drive engaged first in 'ahead', then in 'astern', for intervals of 30 seconds. This will allow time for air to be expelled from the system. After this initial run the engine should be shut down and a check made on the oil level. Top up if necessary. A close inspection should be made of all hose connections to ensure there are no leaks.

With a cable control system operating the control valve, make sure that the correct travel of the operating lever is achieved.

Jet Drives

Instead of a propeller operating in unrestricted conditions, the jet unit is generally an axial-flow turbine with its rotor – or rotors, if a multi-stage design – closely fitting into the casing, which carries bearings at each end of a shaft and a seal at the inboard end. This functions like the inboard gland of a conventional stern-tube. (Figure 8.8)

Maintenance required is negligible. Keeping the water intake

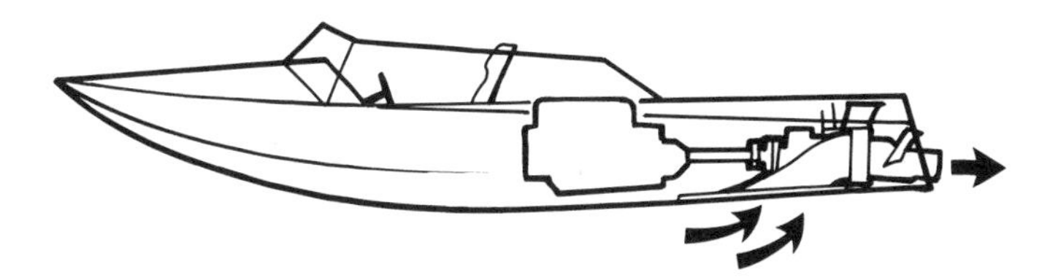

Figure 8.8 *Jet drive*

screen free of weed etc. and lubricating the reverse flap and steering mechanisms is all that is generally required, except for tightening the shaft gland and maintaining a supply of oil or grease according to the type of lubrication. Replacement of the gland packing or sealing rings will eventually be necessary as well as the shaft bearings.

Engine and Transmission Controls

Although the engine throttle and stop controls are not associated with the drive train I have included them with the transmission controls, because a single-lever arrangement operating both systems is the most common type in use today.

Many of the faults initially attributed to the gearbox's failure to engage gear properly or return to the neutral position, as well as inability to obtain maximum engine revs, are caused by a malfunction of the controls. There are several possible reasons for this, and the following notes provided by Teleflex Morse Ltd. give explanations for common service problems and advice on how to maintain push-pull control systems in good conditions. This applies also to steering cables as used with stern-drives and outboard motors.

Push-pull cables cannot be repaired.

In fact, internal lubrication, which is usually the most commonly applied 'cure', may temporarily hide the symptoms of a failing cable, causing a false and dangerous sense of security.

Any indication of faulty control or steering system function calls for an immediate and thorough inspection. If a damaged or poorly functioning cable is found, replace the cable. They are easy to install, relatively inexpensive, and readily available from your distributor.

The following steps are intended to help you check for some of the most common problems. If you encounter a problem that isn't

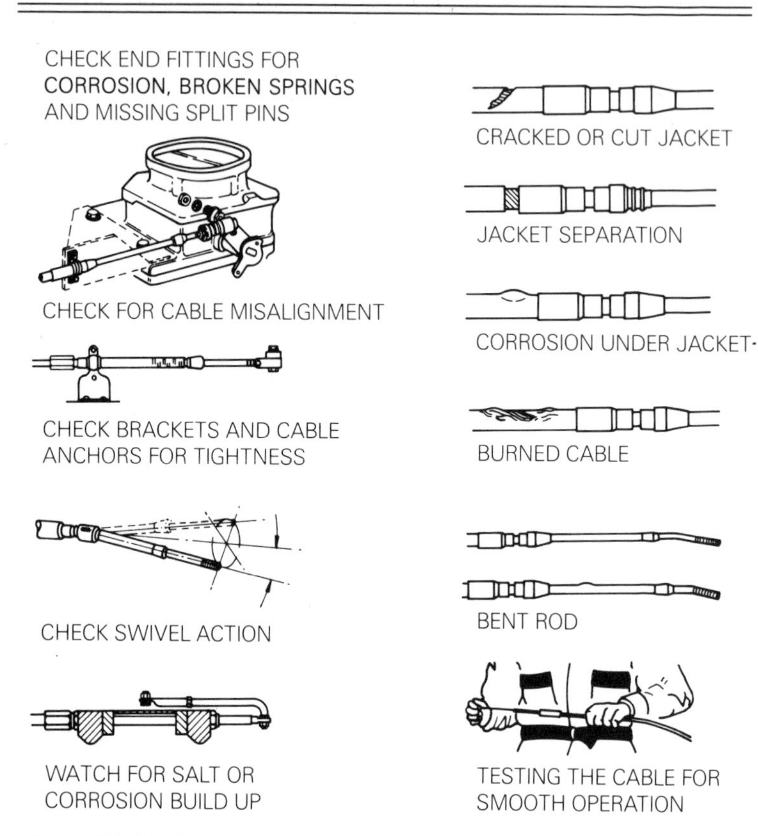

Figure 8.9 *Engine controls. Inspection of push-pull cable systems.*

covered here, refer to the control or steering system installation instructions or contact the manufacturer's service department. (Figure 8.9)

The Visual Inspection

Many of the problems which occur in steering and engine control systems can be spotted just by giving the systems a careful looking-over. Here are some key things to look for:

Cable Attachment

The very first areas to check out are the cable connections at the engine and/or rudder. Make sure the hubs are securely anchored. Check brackets and cable anchors for tightness.

Inspect terminals and end fittings for wear, corrosion, broken springs and missing split pins, and replace if necessary. When making adjustments at threaded connections, be sure that thread engagement is at least equal to 1½ times thread diameter.

Check that all cable seals are in good condition and correctly installed.

Check steering operation to make sure motor turns equally in both directions. The steering cable output end must not touch fuel lines, control cables, electrical wires, etc. when operated in both running and tilt positions.

Control Cable Alignment

Most push-pull cables for marine control applications have a built-in swivel to compensate for minor misalignment, and cable-to-engine connection kits are designed to keep misalignment well within these tolerances.

Excess misalignment will cause poor control cable performance and can contribute to premature cable failure. Inspect the alignment of the control cable ends where they attach to the clutch and throttle levers of the engine.

Corrosion

Corrosion of cable end fittings and/or the connection parts usually calls for replacement of damaged parts.

In addition, build-up of salt or other deposits should be regularly removed. This is particularly critical on through-tilt-tube steering installation and calls for the removal of the cable from the tilt tube to remove deposits from the cable end and tilt tube. Prior to reassembly the output end of the steering cable and the tilt tube should be relubricated with a good grade marine grease such as Keenomax L2. Ensure that with the steering cable disconnected, the engine or rudder moves freely on its bearings.

Cracked or Cut Jacket

The plastic outer covering or jacket of push-pull control and steering cables protects the cable from corrosion. If any cuts or cracks penetrate the jacket anywhere along its length, the cable must be replaced immediately.

While checking for cuts or cracks, also watch for signs of abrasion, which may indicate potential 'wear through' problems. Re-locate the cable to avoid wear points.

Jacket Separation

A key problem area, particularly for older push-pull cables, is the point where the plastic jacket, or outer covering of the cable is joined to the metal end fittings. Any separation or opening at this point can allow corrosion, leading to poor operation and cable failure.

Corrosion under Jacket

Another indication of a problem in this area is any 'swelling' of the conduit. This means that corrosion is taking place within the cable, even though there are no apparent openings in the jacket.

If any of these conditions are found, replace the cable immediately.

Burned or Melted Jacket

While modern push-pull cables can withstand temperatures normally encountered in marine installations, any signs of burning or melting of the outer covering of the cable indicate that the safe heat range of the cable has been exceeded, and the cable should be replaced. When replacing the cable, be sure to route the cable away from the source of the heat, or provide shielding to protect the cable.

Bent Rod or Sleeve

Damage to the rod end of a cable is a common cause of hard operation and cable failure. To check, disconnect the cable from the engine and control ends and visually check the rod end for straightness.

Also check the sleeve portion of the cable end fittings for any deformation, such as a lump or 'bubble' which indicate that the rod has been subjected to side load beyond its design limits.

Another important place to check for a bent rod is at the base of the threads at the end. Bends at this point are not readily apparent unless the attaching terminals, or end fittings, are removed.

Any bend in the rod or sleeve portion of the cable indicates that permanent damage has been done to the cable and calls for immediate replacement.

Performance Checks

Not all of the damage which can occur to a push-pull cable is readily apparent from a visual inspection, so when you are inspecting controls and steering cables, also check out their performance.

Smooth Operation

Before disconnecting cables, operate controls through the entire range. The effort required to move the cables should be constant, with no 'sticky-spots' or hang ups (except, of course, the detents at forward, neutral and reverse).

If sticking, or roughness is encountered, disconnect the control cables from the engine and transmission, and again operate the controls through the entire range. If operation is now smooth, with engine and control disconnected, adjustments are needed at the engine and/or the control head. Consult the manufacturer's instructions.

If operation is still rough after disconnecting cables at engine, disconnect the cables at the control head and stroke the cable by hand through their entire range of movement.

The cable should move smoothly, with a constant, even amount of resistance through its travel. Jerkiness, or sticky spots in operation indicate a problem such as kinked core, calling for cable replacement.

Whenever any part of the control or steering system is disconnected or disassembled, be sure to reassemble and adjust the system exactly according to the manufacturer's instructions. Carefully check the entire system to make sure all fittings and fastenings are properly installed, adjusted and tightened.

Old Age

Very few of the push-pull cables used for marine engine and steering controls truly wear out. But they can. And again, failure can be unexpected.

If the cables on your boat are more than a few years old, or if they've been given rough or continuous use, we suggest replacement. It's an ounce of prevention that can easily be worth more than the pound of cure.

Steering System, Push-Pull Cable Maintenance Service Checklist

Item	*Check for*	*Service*
Bolts, nuts & fasteners	Looseness, corrosion, damage or wear.	Tighter to manufacturer's specs. Replace corroded, damaged or worn parts.
Helm assembly	Check all bolts, nuts and fasteners for tightness.	Tighten as necessary.
Steering wheel	Looseness, cracks around hub and base of spokes.	Replace wheel if cracked. Tighten retainer nut and check key if wheel is loose.
Cable mounting bracket/fittings	Corrosion, damage or wear. All bolts, nuts and fasteners for tightness.	Tighten as necessary. Replace corroded, damaged or worn parts.
Engine steering fittings	Looseness, corrosion, damage or wear.	Tighten to manufacturer's specs. Replace corroded, damaged or worn parts.
Cable conduit	Cracks in outer cover, abrasion of cover, kinks or sharp bends.	Replace cable. Use only genuine Teleflex Morse Cables.
Cable ram	Proper lubrication, bent or distorted.	Clean with mild solvent and relubricate, replace cable if ram is bent.
Engine trimtab	Corrosion, damage.	Replace damaged or corroded tab.
All moving parts	Cleanliness and lubrication. Freedom of movement.	Clean and lubricate all moving parts with a good grade of marine grease. Clean and lubricate every three months in fresh water areas and every month in salt water areas. Use mild solvents such as paraffin for cleaning.

Control System Maintenance Service Checklist

Item	*Check for*	*Service*
Remote control head	Cleanliness, corrosion, loose mounting screws or other parts.	Remove dirt, grime or corrosion. Wax with automotive wax. Tighten loose parts as necessary – spray control mechanism (behind mounting panel) with moisture displacing lubricant such as WD-40, or equivalent.
Cable conduit	Cracked or abrasion of outer cover. Kinked or bent cable.	Replace cracked, kinked or bent cable. If abrasion has not worn through outer cover, cable is still serviceable but cable must be protected from further abrasion.
Cable ends & connection fittings	Loose, worn, or damaged fittings. Loose cable brackets. Corrosion.	Replace corroded, worn or damaged parts. Tighter loose parts as necessary. Spray cable ends and fittings with a moisture displacing lubricant such as WD-40, or equivalent.

Service

As we've said, control and steering cables are not repairable, and if a cable starts to fail, no amount of service will offer more than a very temporary improvement.

But that doesn't mean that preventive service isn't necessary!

Regularly check connections, cable anchor points, etc., for tightness.

Lubricate exposed moving parts of cable to engine connection fittings to reduce friction and wear, and to inhibit corrosion.

Keep throttle clutch, and steering mechanisms adjusted to the manufacturer's specifications.

9 Propellers

Lengthy books have been written on the subject of the design, manufacture and testing of propellers, but the owner of a power-boat or auxiliary-engined yacht will not need to study the subject in great depth unless, of course, his appetite is whetted by the contents of the next few pages.

It does seem curious that great importance is rightly attached to the choice of an engine and its installation in the boat, but that the propeller is liable to get scant attention in pleasure-craft, although the only purpose of the 'machinery' is to rotate the propeller so that this can exert thrust in an aft direction to move the boat forward. The propeller also needs to change the direction of this thrust to oppose the forward movement of the craft and thereby apply the brakes.

Whether the engine power is used to the best advantage is largely dependent on the efficiency of the propeller. A badly matched 'wheel' will waste power and hence fuel, as well as reducing the boat speed below its capability. The importance of propeller efficiency increases with more powerful engines; a ten horsepower auxiliary engine wasting one horsepower because of a poor propeller may make no discernible difference to the motoring performance of an auxiliary yacht when it is fitted with a 'good' propeller. However, the same 10 per cent power wastage in a 1000 bhp trawler installation could necessitate running the engine at a higher throttle setting to achieve the same boat speed or trawling capability, with significant fuel wastage as a consequence. The selection of propellers for large commercial craft is treated very seriously, and nowadays computers are used to assist in producing the most efficient blade shapes as well as matching the diameter and pitch to the hull and engine characteristics.

For high-speed pleasure-craft the propeller is designed to achieve maximum boat speed with little consideration for engine fuel economy, although this is becoming more important with today's high fuel costs. However, for long passages the amount of fuel to be carried has to be taken into account, so that economical operation becomes necessary and a compromise between speed and range is often required.

The propeller used by a tug is required to develop maximum thrust, for the vessel to have the best towing or pushing performance. Its design is therefore directed towards this objective, with vessel speed of secondary importance.

In order to achieve maximum possible propulsive efficiency the propeller has to be considered at an early stage in the design of a boat, so that proper provision is made for it and it does not have to be squeezed into too small a space for efficient operation. The engine power necessary to propel the boat and the rotational speed of the propeller shaft have also to be decided. The latter will determine whether the engine can have a direct-drive transmission or whether it requires reduction gearing. Fast boats may have direct drive – racing craft sometimes use a step-up ratio. Cruising boats usually have 1.5:1 or 2:1 reduction, while work-boats may need a 3:1 or larger reduction ratio, depending largely on the speed of the boat and the engine-rated speed.

Propeller Specifications

Propellers come in many shapes as well as sizes and are manufactured in various materials chosen according to the type of duty, the design of the blades and considerations of cost. Most propellers are cast in a mould, so the material and the shape have to be suitable for this process. Manganese bronze and aluminium bronze are the most common materials. Cast iron, steel, stainless steel, aluminium alloys and plastics are also used.

The measurements of a propeller defining its specification relate to diameter, pitch, blade area and number of blades. The first and last are obvious; pitch can be defined as the distance moved forward during one revolution if the propeller was working in a solid material rather than water, where it will 'slip'. The amount of slip, which is really how much the propeller does not move forward in one revolution, is usually expressed as a percentage.

Some people are concerned that their propeller may have

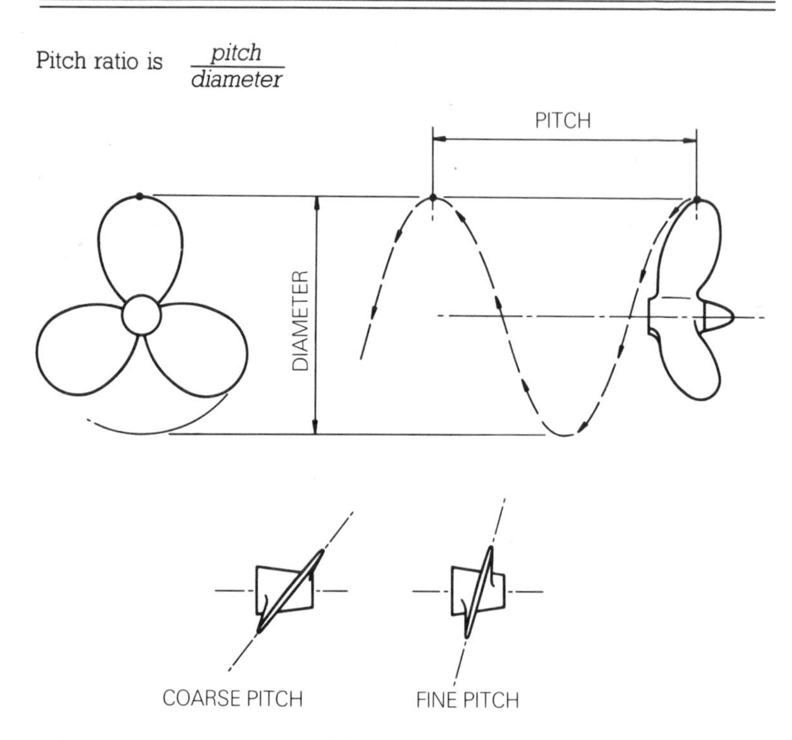

Figure 9.1 *Propeller pitch. Pitch is measured linearly.*

excessive slip and think that it should be eliminated to achieve maximum efficiency. In fact it is necessary to have a certain amount of slip in order to produce thrust; in practice this varies from about 10 per cent for high speed craft to 45 per cent for slow work-boats. Figure 9.1. The blade surface area which is usually specified as a percentage D.A.R. (disc area ratio) or B.A.R. (blade area ratio) is the ratio between the total area of the blades and the circle described by the blade tips. (Figure 9.1)

A further variant in the specification is the shape of the blade and its cross-section area. Sturdy propellers with simple blade shape and relatively thick sections are used for heavy duty work where abrasion and damage are liable to occur. Fine blades with slim sections of specially designed shape to cut cleanly though the water with minimum disturbance are used for high speed planing boats. Even finer section blades of stronger materials such as stainless steel are used for racing craft; an 'aerofoil' section as used for aircraft wings is sometimes employed. Special-purpose blade shapes have been developed for specific purposes, for example the 'weedless' propeller generally available in two- or three-blade versions and the high-efficiency blades of the surface-piercing propeller. Both types are illustrated in Figure 9.2.

Figure 9.2 *Propeller types*

1 Turbine
2 Equipoise
3 Two-bladed
4 Two-bladed folding
5 Weedless
6 Surface-piercing (super cavitating)

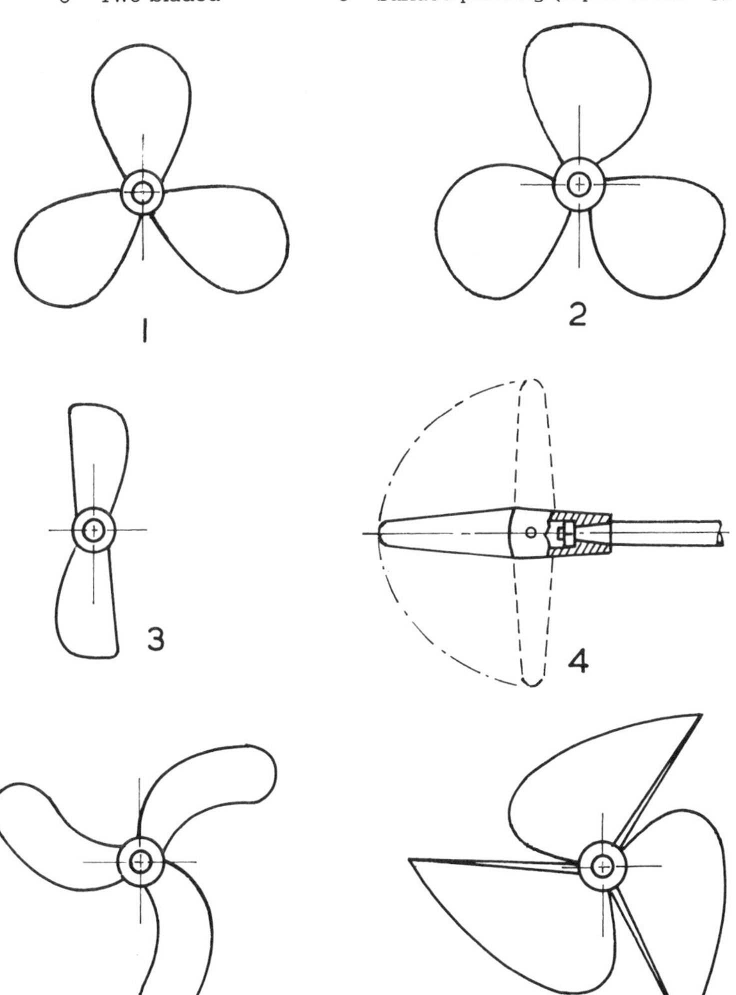

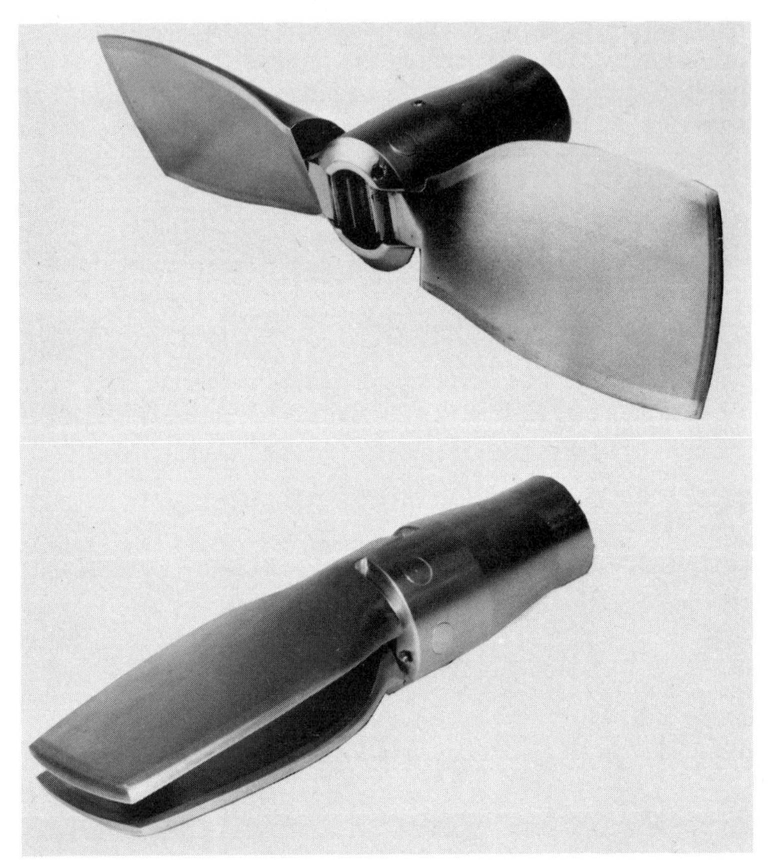

Figure 9.3 *Folding propeller*

The number of blades is generally two or three and less commonly four or even five. Except when two blades are used because the propeller has to be locked vertically behind the sternpost when sailing, or because the propeller is of the folding variety, the number of blades depends on the blade surface area required. Too much surface area distributed among only three blades would make the blade shape inefficient so four are used. Four blades are also sometimes used when there is a restriction in the diameter allowable.

The propeller manufacturer will recommend the style of propeller and the number of blades as well as the diameter and pitch for a particular boat and engine combination. Smoother running characteristics are attributed to four-bladed propellers, which help to reduce stern vibration; reversing performance and manoeuvrability are also improved because the four blades have a better 'grip' on the water.

'Cupped' propellers are produced by Michigan Wheel to improve the performance of high-speed craft; the cupping is applied to the trailing edge of the blades to increase the jet stream volume and reduce the tendency for cavitation and slippage occurring at high rotating speeds.

The folding propeller illustrated in Figure 9.3 is made by Teignbridge Engineering. The geometry of the blades is such that they are held in the folded position for sailing; when the motor turns the shaft, centrifugal force opens the blades to the working position in ahead or astern rotating directions.

Cavitation

This phenomenon can affect propellers of any type applied to any type of boat. It appears first as a kind of rash of pock marks on the back surface of the blades, i.e. the side on which the thrust is developed when the boat is moving forward. This grows into patches of erosion which can eat into the blade, weaken it and cause failure.

The cause of cavitation is excessively high pressure on the blades, caused by a propeller rotating too fast or there being insufficient blade surface area. When the thrust pressure is too great, small cavities are formed in the water adjacent to the blades. The water boils in the cavities because the pressure is very low, and as each cavity collapses like a bubble bursting, tremendous force is exerted on a very small area, eroding the blade surface. Factors other than high blade pressure loading can lead to cavitation – for example, badly-faired hull sections that prevent smooth water flow into the propeller, and blades that are too close to the surface and draw air into the propeller.

There is a calculation which can be applied to a propeller design to check the probability of cavitation. I say 'probability' because widely varying propeller designs and different materials, as well as the effect of the installation, can make a difference to the actual loading which will cause cavitation in a particular circumstance.

The calculation first determines the thrust developed by the propeller, then the blade pressure, i.e. thrust per unit area, which is then compared with the permissible loading shown below the line on the chart. This is really in the big propeller league where measurements are in feet and inches (if we stay with the imperial standards still favoured by many builders and owners).

1. Calculate slip:

$$\text{Slip} = \frac{(\text{Pitch in feet} \times \text{propshaft revs per minute}) - (101.3 \times \text{boat speed in knots})}{\text{Pitch in feet} \times \text{propshaft revs per minute}}$$

2. Calculate propeller thrust:

$$\text{Thrust} = C \times D^4 \times \left(\frac{\text{shaft revs per min}}{100}\right)^2 \times P(P \times 21) \times S$$

where
C = Blade constant (0.11 where blade/area ratio is 0.5)
D = Diameter of propeller in feet
P = Pitch ratio i.e. $\frac{\text{pitch}}{\text{diameter}}$
S = Slip percentage

The thrust figure calculated will be in lb./sq. in. units and can be plotted on the chart, Figure 9.4. If it is well below the line the propeller is likely to be free of cavitation unless the installation is poor, as described earlier. Note how the permissible blade loading increases with higher boat speeds.

Apart from running your engine at reduced speed so that the power absorbed by the propeller (and hence the thrust developed) is reduced, there is little that you can do to halt the cavitation process short of exchanging the propeller for one with a greater blade surface area, to reduce the surface pressure. This might entail using a four-bladed model instead of three, or a different type of propeller with higher B.A.R.

Propeller Measurement

The diameter and pitch of a propeller are often stamped on the boss. If not, or if you suspect that it may have been modified, you can easily take the necessary measurements yourself.

Start by placing the propeller on a sheet of paper laid on a flat surface, with a peg in the centre from which you can measure the radius of the blades and hence the diameter. The blade area can be measured by making a template on squared paper of one blade,

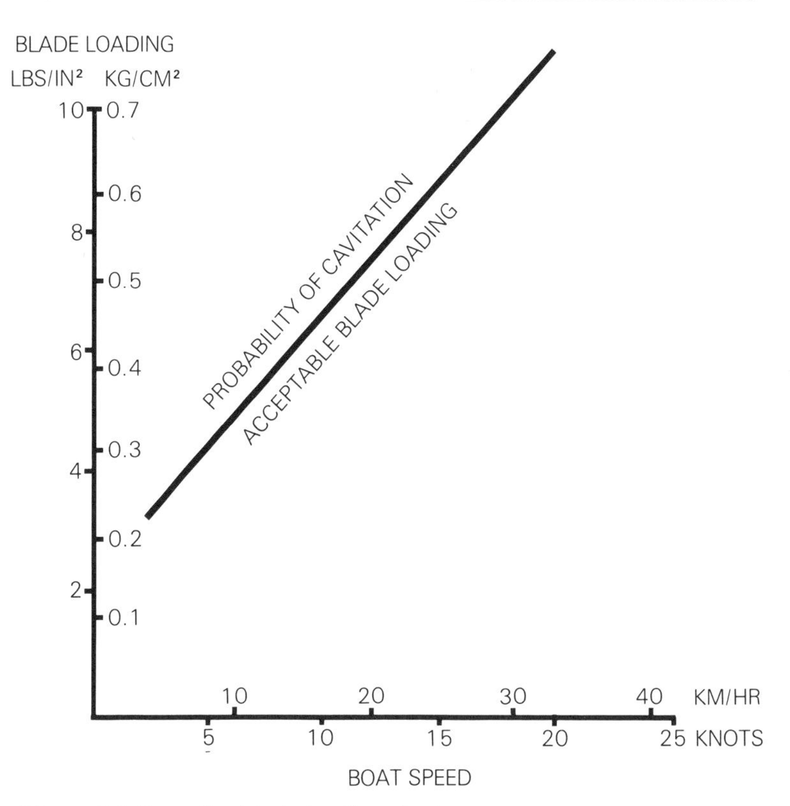

Figure 9.4 *Propeller blade loading chart*

then counting the squares, making allowance for part-squares where the edges cut through the lines.

Pitch measurement is not difficult to calculate. Working from the centre of the peg, scribe a line on the surface of a blade across the edges at approximately two-thirds of the radius of the tip. Using a square, project the two points where the scribed line cuts the edges of the blade down onto your sheet of paper so that, with the propeller removed, you can measure the angle ø indicated in Figure 9.5. Before removing the propeller, measure the two distances x and y of the blade edges above the paper surface. The pitch will be $(x - y) \times \frac{360.}{ø}$

Most propellers are *constant pitch*, i.e. the pitch is the same at any radius from the boss to the tips, and the twist of the blades forms a helix. Working in a solid the pitch would have to be constant, but in

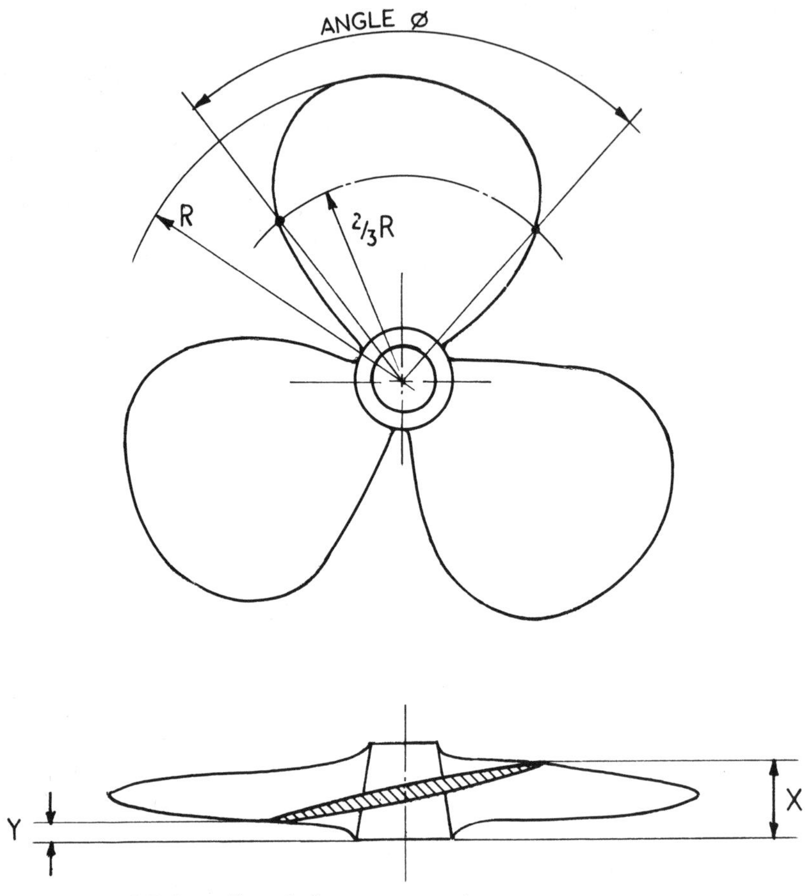

Figure 9.5 *Propeller pitch measurement*

water this is not so and the pitch is sometimes varied from boss to tip to improve propulsive efficiency. To check a propeller with varying pitch we have to measure it at the point where it is nominal – usually about two-thirds of the radius.

The accuracy of a propeller can be checked by measuring the radius of all the blade tips and also their pitch. A pitch variation in (x – y) of only 1 mm. (0.040 in.) with angle ø of 60° would indicate a pitch variation of 6 mm. (¼ in.) *at the points of measurement.*

The pitch of propellers can be checked at all positions on a blade by means of pitch blocks. The distortion caused by damage can be quickly seen, and propellers of ductile material can be hammered back to shape until they fit the pitch block profile. It is perhaps surprising how very badly damaged propellers can be built up with the original material hammered to shape, trimmed to the blade profile and balanced, to look as good as new. Propeller repair is a highly skilled procedure which is not recommended for the amateur.

Propeller Selection

The same combination of engine, gearbox and prop-shaft could be fitted with many different propeller specifications i.e. variations of diameter, pitch and blade design, all of which could be chosen to load the engine to the same extent, yet because of their different degrees of efficiency each would drive the same boat at a different speed. For example, a 20 in. diameter with 18 in. pitch might absorb 100 horsepower at the same shaft speed as a 19 in. × 19 in. propeller, yet, especially in a light planing hull, there could be a different maximum speed of a knot or so.

But maximum speed is not everything – a propeller giving a slightly lower maximum speed could develop more thrust when the engine is accelerating and bring the boat more quickly onto the plane. Extra thrust could be important when towing skiers. A fully-loaded boat would be better served by a slightly 'smaller' propeller which would not absorb the full 100 bhp when the boat was lightly laden.

Manoeuvrability is a further consideration. In the example given, a 21 in. × 17 in. propeller could be less efficient than the other two as regards obtaining maximum boat speed and acceleration, but the larger diameter blades would get a better 'grip on the water' when going astern.

Apart from the examples given, there are other factors to be taken into account when selecting a propeller. Is the boat to be used in fresh water or salt? A slightly larger propeller could be necessary in fresh water because of its reduced density and hence resistance. Is the boat to be used in very hot or very cold climates? This will affect the engine power which is less when hot air is drawn into the cylinders because of the lower weight of oxygen, although turbo-charged engines are not so much affected as naturally-aspirated ones.

Especially for high speed craft some trial and error is necessary before the final propeller is decided on and it is good practice to carry out extended trials under differently loaded conditions. Before starting, however, some idea of a theoretical size must be obtained, and there are several methods of calculating a propeller by means of specially-prepared slide rules, tables, charts and computer programs. The data on which these are all based come from tank tests conducted on 'model' propellers of the same diameter but different pitch and blade design which will run at different speeds with measurements taken of the power absorbed. Data obtained from experiments carried out by the Netherlands Ship Model Basin is used by many European manufacturers of boat engine propellers.

Whatever method is used for propeller calculation it is necessary to have accurate information on the engine power output, reduction gear ratio, power loss in the gearbox and (which is often the most difficult to predict in advance) the likely boat speed with the power available.

Chapter One explained the different ways of expressing the power output of the engine. In quoting the power available to the propeller we can take the *net* power at the flywheel and deduct the gearbox losses, then the shaft-bearing loss. Alternatively we take the *shaft* horsepower, i.e. power at the gearbox output flange, and then take off 1 per cent for each shaft bearing.

The boat speed is best arrived at by actual results with the same type of craft, or at least the same hull, using the same engine power and with identical displacement. Some designs provide a table of boat speed against power and hull weight, in which case you simply read off the appropriate figure. But unless your boat is one of a production class (in which case the propeller size will already have been determined) you are unlikely to find your precise combination of hull design) power, and all-up displacement in the boat. However, if you are providing data for a propeller manufacturer and you can quote a known boat speed for different known (not estimated) power and weight figures, the speed for your boat can very soon be arrived at.

One of the most common mistakes in predicting boat speed is to underestimate the displacement. Even if the performance was satisfactory on initial trials, the almost inevitable addition of cruising gear and the completion of the 'basic' boat specification can add 10 per cent or more to the weight. This might not matter much in a cruising auxiliary yacht, but in a fast planing cruiser it could be crucial to getting up on the plane.

Installation

As mentioned earlier, for best results the propeller must be located so that the blades receive an unimpeded flow of water and the jet stream aft of the propeller is similarly free of restrictions. The radial clearance of the blades from the hull should be a minimum of 15 per cent of the propeller diameter – not only to obtain optimum performance but also because insufficient clearance can cause vibration. Vibration may also be caused by the propeller being out of balance or having a significant blade-to-blade pitch variation. A bent shaft or worn bearings may also cause vibration.

The position of the propeller boss in relation to the aft bearing of the stern-tube or 'P' bracket etc. should be sufficient to allow for axial movement of the propeller-shaft, which can be about 6 mm. (¼ in.) when clearance in the propeller-shaft bearings and deflection of the flexible couplings allows the shaft to move forwards under the action of the propeller thrust. With larger engines a force of several tons is exerted by the propeller as it starts to push the boat forward; this thrust is transmitted to a thrust bearing which is generally located in the gearbox. The engine mountings transfer the thrust to the engine beds, which is one reason why these have to be strongly constructed and properly tied to the hull.

Stern-gear

In relation to the cost of the power unit, the simple arrangement of a propeller-shaft and stern-tube, with its bearings and possibly a 'P' or 'A' bracket behind the propeller, may seem expensive when added to the cost of the propeller. This area can become vulnerable to cost-cutting, and so undersized propeller-shafts, a smaller diameter propeller than is necessary, and unsuitable materials may result when building down to a price. Fortunately most of these shortcomings become obvious when the boat is in service, and such practice – especially for larger sea-going craft – is rare.

Some of the problems encountered in service, and good design practices to overcome them, are given below.

Propeller

It is important to be able to remove the propeller from the shaft end and it is not good practice to have to do this with a hammer and chisel. Two threaded holes in the boss, into which substantial extractor bolts can be screwed, are ideal, especially for larger propellers. The shaft taper should accurately match that in the propeller hub. A locking device in the propeller retaining nut is desirable and the design of this should be appropriate to the size and duty of the propeller.

Adequate blade tip clearance will help to minimize damage caused when pieces of wood etc. become trapped and by the effects of the sudden stoppage on the gearbox or couplings, as well as on the propeller itself and its shafting. Too much overhang from the aft support bearing should be avoided in the interest of minimizing shaft distortion when the propeller motion is suddenly arrested. Flexible couplings, especially when fitted in pairs, will help to minimize damage.

Propeller-shaft

For economy, or to facilitate assembly and removal, an intermediate shaft of cheaper material such as mild steel is sometimes used – the tailshaft usually being made of manganese bronze, stainless steel or Monel alloy. Experience has shown that stainless steel of good quality, e.g. EN58J grade, is necessary for the tailshaft.

To avoid corrosion, sharp corners on the shaft must be avoided, and because of the possibility of electrolytic action caused by the different materials of propeller, shaft and stern-tube etc., anodic protection as described in Chapter Ten should be provided and maintained in good order, particularly the earthing of the anodic protection system. Attention should also be paid to the earthing of electronic equipment etc.

Good practice in the manufacture of propeller-shafts ensures that the bar material is carefully straightened before the ends are machined. Shafting should then be stored on end to avoid deformation.

Propeller-shafts become worn in the areas of the bearings, and eventually this *wasting* of the shaft will cause failure because of insufficient strength to transmit the engine power.

An ingenious way of doubling the life of a shaft is to machine the same tapered end in the coupling as in the propeller boss. When worn, the shaft is removed and replaced the other way round. Of course, this assumes that the bearings are not positioned at the same distance from each end of the shaft – an unlikely requirement. Building up the shaft by metal spraying, welding or 'Fescalizing' is another possibility.

Shafts which have been damaged so that they are bent or twisted can rarely be corrected satisfactorily. However, a simple bow or 'S' in the shaft length can usually be straightened by a repair shop or the stern-gear manufacturer.

To check a shaft, it should be supported in vee-blocks or rollers positioned where the stern-tube bearings are located. By carefully turning the shaft a few degrees at a time and taking measurements at several points along the shaft, the actual deviation from the true shape can be plotted. This information allows a skilled operator to apply pressure from hydraulic jacks, with the shaft clamped into a fixture such as a lathe bed, to bring it back close to its true shape.

Couplings

Close control of accuracy in machining half-couplings so that the shafts run concentric to the gearbox flange is very important. The coupling design should incorporate a *spigot*, fitted bolts or dowels. Simply bolting the flanges together is not good enough, whether the couplings are solid or flexible.

10 The Winter lay-up

'Winterizing'

Whether you decide to lay the engine up for the winter, or leave it in runnable condition but take the necessary precautions against frost damage, depends on where you keep it (on shore or on the mooring), how long it will be out of use, and whether you intend to pay it regular visits.

One thing you should not do if the boat remains afloat is to keep running the engine in neutral gear. This applies in fact at any time of year, because engines do not benefit from long periods of light-load running. Especially when they are new, prolonged running at low to medium speed will cause the cylinder liner surfaces to become glazed, so that the piston rings cannot bed-in properly and will allow lubricating-oil to pass up into the combustion area. Oil will burn with its characteristic blue colour in the exhaust and your consumption will be higher than it should be. The only remedy approved by most engine-manufacturers is to remove the cylinder-head and use a 'glaze-busting' tool, which looks rather like a bunch of wizened grapes – lead pellets on wire stalks. (Plate 10.1)

Frost Protection

Whether laying up the engine or not, you must guard against frost damage in the cooling system. With heat-exchanger cooling this necessitates draining the sea-water circuit, because although sea water freezes at a lower temperature than fresh water you can still get ice in the sea-water pipework, heat-exchanger, coolers and water-pump etc. Make sure that all the drain plugs or cocks are opened so that water that collects, for example, in the bottom of

Plate 10.1 *'Glaze Busting'*

pipes and coolers is drained out. Poke out drain holes with a piece of wire if water does not run freely. Sometimes it is necessary to slacken off pipe connections and remove hoses to drain the system. Remove the rubber impeller from the water-pump, tie it to the pump by a piece of string and refit the cover and screws. Run the engine slowly for a minute to pump water through the circuit.

With the fresh water circuit you can either drain or use anti-freeze. Unless you are going to start your season in freezing weather, there is no point in using anti-freeze – unless you have an engine that cannot be drained completely. In this case use a type of anti-freeze recommended by the engine manufacturer. The British Standard for anti-freeze is BS 3151. This incorporates a corrosion inhibitor. A strength of 25 per cent will protect down to –12°C (10°F).

Fuel System

Fuel should not be left in the engine all winter if it is being laid up. Petrol will leave gummy deposits and diesel fuel can precipitate wax at low temperature. Arrange your final trips so that there is little fuel to drain out of the system; the engine as well as the tank and filters should be drained off.

Diesel engines should have a preservative oil such as Shell Fusus 'A' fed into the system, either by putting a gallon or so into the main tank or by introducing a smaller quantity into the pipeline at a flexible connection so that it can be pumped through the system by the priming pump, remove bleed screws to permit the oil to flow as explained in Chapter Three. Turn the engine over or run for a few minutes to get the fuel to fill the injection pump and high pressure lines.

Lubricating Oil

Seal the oil filler cap and breather connection with adhesive tape. If you have not had a recent oil change it is a good plan to do this before the winter lay-up. Change the filter element, clean out the breather and any other items specified for the lubrication system. Fill the engine with fresh oil and run the engine for a few minutes. This will guard against corrosive acids attacking the oil-ways during lay-up.

Electrical System

Remove the battery if possible, so that you can recharge every month or so. Alternatively remove the leads. Clean the terminals and smear lightly with petroleum jelly. Spray the top of the battery with water repellent if it is left in the boat.

Starter and Alternator

Clean terminals and smear with petroleum jelly. If you decide to remove electrical components to avoid corrosion, label the harness connecting terminals.

Cylinder Bores

Remove the injectors or spark-plugs and spray a small quantity of lubricating-oil into each cylinder. Turn the engine over and replace the injectors/plugs.

Exhaust Pipe

Seal the pipe off at the transom outlet, or remove at the manifold joint and seal the manifold with polythene or similar waterproof material.

Air-cleaner

Seal the air inlet connections or remove and seal off the induction manifold or carburettor. Squirt, or better, spray a mist of lubricating-oil into the manifold. (Figure 10.1)

Jobs for the Winter

While the boat was in use during the season you probably decided that several jobs should be done while it was laid up for the winter. It is suggested that you make out a list, and if you intend to do the work yourself, or arrange for a specialist to work for you, it is best to collect your components or materials in good time. This applies whether you are intending to work on the boat itself or on the engine and its installation.

There are two separate categories of jobs to be done: repairs to rectify faults which developed during the season, when you may have got by with a temporary 'fix' (to the engine pipework for example), and improvements to the 'machinery' and its installation.

Having carried out any necessary repairs you can then make improvements which would enhance your safety, well-being and general enjoyment of next season's boating.

It goes without saying that standards of engine installation in pleasure-craft and work-boats vary considerably. It may be mainly a question of how much money the boat-builder has allocated to the

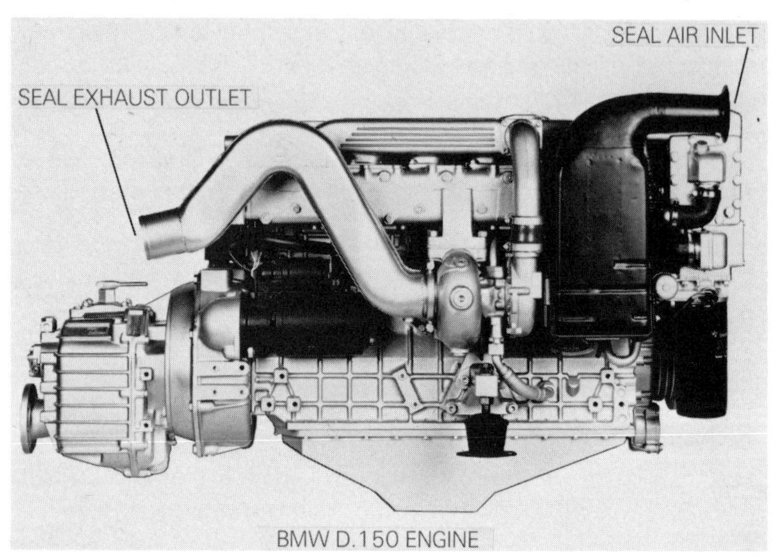

Seal exhaust and induction systems
Drain seawater and remove pump impellor
Drain fresh water
Seal lub. oil filler and breather
Inhibit fuel system
Clean electrical terminals, smear with petroleum jelly
Check manfacturer's specific recommendations

Figure 10.1 Winterizing

engine and its associated material. Even so, using unsuitable items such as badly-matched engine mountings and tailshaft couplings or inferior soundproofing materials can result in a poor installation, whereas the same amount of money spent wisely could give a quieter, smoother-seeming engine. Most engine-manufacturers supply – or at least recommend – properly-matched engine installation material; the bigger companies spend considerable time and money in experimenting with installation fittings such as flexible mountings and couplings, for example, so it is best to use what they recommend rather than 'Brand X' offered by the local boatyard.

Some of the principal ways in which you may be able to improve your machinery installation are dealt with in the following pages.

Soundproofing

Although petrol engines benefit from adequate soundproofing –

particularly when operating at high speeds – it is particularly important for diesel engines. Figure 10.2 shows an installation with many faults, causing noise and vibration to be transmitted to the wheelbase and accommodation areas. Figure 10.3 has all these problems rectified. The effect of such a transformation has to be experienced to be believed.

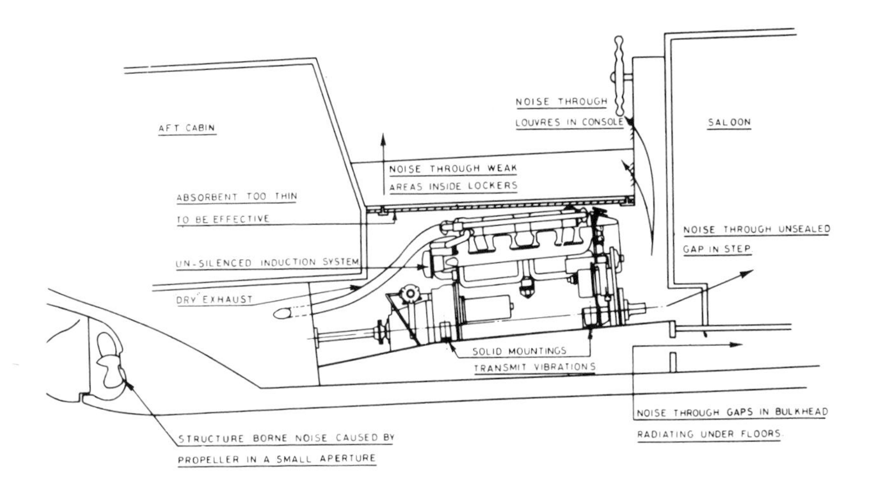

Figure 10.2 *A potentially noisy installation*

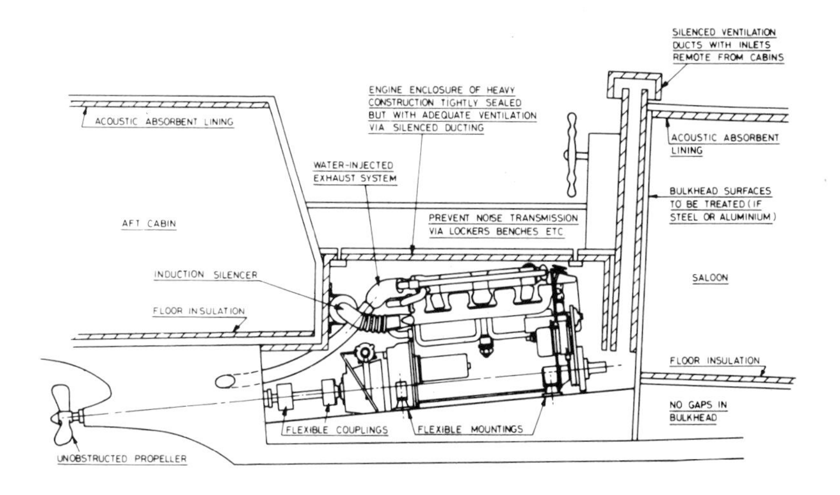

Figure 10.3 *A quiet installation*

The first fault illustrated is noise paths caused by holes through the engine enclosure. Often over-generous clearance holes are provided for pipes, cables, controls etc. The remedy is to use grommets, foam rubber or sealants sold for this purpose. It can be difficult to spot all the holes; a light inside the engine compartment may reveal corners where there are gaps in the panelling and also clearance around the pipework.

Soundproofing material should be used on all possible surfaces in the compartment – under the deckhead including the removable panels (which should be bedded onto rubber strips), on bulkheads, and against the sides of the hull. The best product with good sound-insulation properties consists of a sheet of lead or other dense material with a resilient layer separating it from the engine compartment surfaces. On the other side should be 25 to 50 mm. of absorbent material, i.e. foam, or glass fibre mat with a protective skin of PVC, for example, on the exposed surface. Metallic surfaces such as tanks or bulkheads, where it is difficult to apply rigid sheets of material, can be treated by gluing on absorbent foam with a PVC surface. Materials used for soundproofing should have flame-retarding properties.

In the wheel-house and the adjoining accommodation areas, a soundproofed head-lining should be used as well as floor coverings.

The engine must of course be supported on flexible mountings with a flexible coupling or couplings in the shaft line, unless the stern-tube has a flexible inboard gland.

With such soundproofing installed it is possible that the engine exhaust and air inlet become audible. A water-injected exhaust is generally an effective noise-suppressing device, although further improvement can be made with a silencer. Make sure that the silencer is large enough for the engine, so that it does not cause a restriction to the exhaust gas. Engine-builders quote the maximum 'back pressure' measured close to the engine exhaust manifold by means of a manometer. A pressure of 76 mm. (3 in.) of mercury is quoted by Perkins, for example.

If the engine is not fitted with an air silencer, i.e. the intake is a simple gauze screen, then a silencer can be introduced with beneficial results. This is either fitted to the engine or mounted from a convenient location in the engine compartment with a flexible duct to the inlet manifold. (Figure 10.4)

The ducting into the engine compartment assumes greater importance when you have blocked up all the holes as well as put in

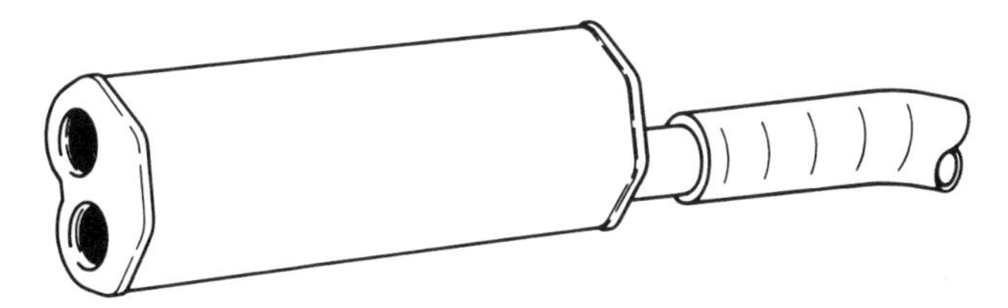

Figure 10.4 *Air silencer. Remote mounted on bulkhead, etc with ducting to engine air inlet.*

extra sound insulation, which will retain heat in the compartment. Check that the size of the ducting meets the engine-builder's specification. If you don't know what is specified there is a simple test that you can carry out when the boat is in the water. Start the engine(s) and run up to full throttle. Watching the tacho carefully, get someone to lift the engine hatch. If the engine speed increases you have a restriction in the ducting, and the ducting will need to be increased in size. You may need to do this test under way, depending on the effect of the engine speed governor.

If you have to replace the ducting you can use a sound-deadened material and incorporate bends to trap the engine noise.

Propeller Noise

This is an uncommon problem on 'production' boats but is not unknown on custom-built jobs where insufficient clearance was allowed between the underside of the hull and the propeller-blade tips. Reducing the diameter of the propeller and increasing the pitch may give the same engine loading but will usually reduce efficiency (i.e. speed or fuel economy) as well as reducing the 'grip on the water' and thus manoeuvrability – especially when going astern.

Some experiments with four- or even five-blade props may bear fruit, but obviously it is a costly job to buy propellers on the chance of an improvement; better to borrow some second-hand props (even if the pitch is an inch or so out), and you may be encouraged by a successful experiment with this makeshift to order a replacement of the same type.

Service Accessibility

While you were doing your preventive maintenance (dipping the oil, changing the filters, etc.) you may have cursed the engine-manufacturer or boat-builder for locating the oil filler, for example, under the deckhead, making access difficult. A small hole cut in the wheel-house floor may make the job much easier to perform. Make sure the cover fits snugly and has its quota of soundproofing material.

There is sometimes scope for moving fuel shut-off cocks to a more accessible position in the pipeline, or fitting an extra cock where you can reach it more easily.

Fit a drip-tray under the engine if this item is missing and if you can get one into position. Copper is a good material, but don't let it touch the aluminium sump or you will have a corrosion problem if there is bilge water in the vicinity.

Engine Mountings

Earlier in the chapter I mentioned the desirability of flexible engine mountings. Any type which will support the weight of the engine will reduce the transmission of vibration to the structure of the boat, but if your flexible mountings are 'tired' so that the rubber is deflecting too much they should be replaced. In fact, unless the rubber has been attacked by fuel or lub-oil leakage, the mountings may have been overstressed simply because they are not adequate for the engine weight. Check what type are recommended by the engine supplier and install this correct type, which will have been selected to give the best isolation for your engine. Most mounting suppliers will recommend a model but this may not in fact be the optimum one, which will be a compromise between smooth running at your cruising speed and reasonable isolation at any 'critical' speeds in the range.

Best isolation is often obtained with considerable engine deflection, but this would mean allowing the engine to roll about and it is not usually feasible to mount a marine engine so that it has the freedom of movement of a car engine. I have mounted small four-cylinder diesel engines in a light aluminium boat on three-point mountings with considerable deflection. Isolation was excellent but scarcely practicable for sea-going conditions in rough weather. A typical mounting is illustrated in Figure 10.5.

Flexible Couplings

With the engine on flexible mounts you need a flexible propshaft coupling unless the shaft runs through a 'log' which allows deflection of the inboard end. With bearings at both ends of a conventional stern-tube, a flexible coupling is a good idea even if the engine is

Figure 10.5 *Flexible engine mounting. Silent bloc pedestal type with height adjustment.*

solidly mounted. Two flexible couplings may seem an unnecessary complication but will often isolate vibration from the propeller-shaft. If the distance from the gearbox coupling to the stern-tube gland is short, a double coupling assembly may be needed.

The boat is pushed along by propeller thrust, which passes from the shaft through the coupling(s) to the gearbox flange, and then via the gearbox shaft thrust bearing through the gearbox casing to the engine – and finally to the engine-bearers via the flexible mountings. In order to transmit this thrust in both directions the flexible couplings as well as the mountings have to be *axially* stiff although allowing *radial* and *angular* deflection. This means that they must be purpose-made. Of course if you have a thrust block at the end of the stern-tube – uncommon in modern pleasure-craft – you don't push the boat along via the engine, so your mountings and coupling do not have to transmit thrust.

I have gone into a little detail here so that you can judge whether your coupling and mountings are adequate. If you can feel vibration from the prop-shaft it is very possible that the mounting and coupling set-up is not an ideal one. (Figure 10.6)

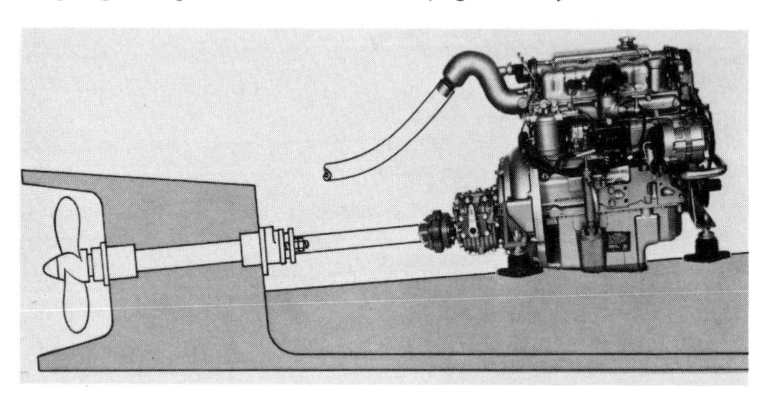

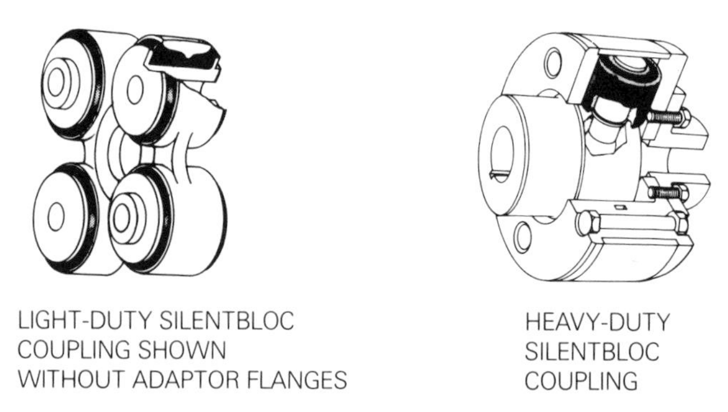

Figure 10.6 *Flexible coupling for propeller shaft. Perkins 4.108 engine.*

Prop-shaft Alignment

Whether you have renewed your mountings or not it is necessary to check the alignment of the engine to the prop-shaft. If you run with the engine out of line there is a risk that the gearbox shaft bearings will fail, or that the stern-tube bearings or the prop-shaft will rapidly wear. You will also waste fuel because of the unnecessary friction caused.

Check the alignment with the boat in the water, to allow for hull distortion; with new mountings it is best to leave the job as long as possible to let the rubber settle under the engine weight. The classic method of lining up is by means of feeler-gauges to measure the gap between the gearbox coupling and the mounting flange, using the adjusters on the mountings or inserting shims underneath the mounts until the two halves of the coupling are in line and parallel.

With a flexible coupling this is not quite so easy to do, and a good practice is to remove the coupling (and if necessary the propeller) so that you can bring the metal part of it up to the gearbox flange. It is not quite so important to have a shaft fitted with flexible coupling dead in line, but excessive misalignment will cause unnecessary bearing wear. (Figure 10.7)

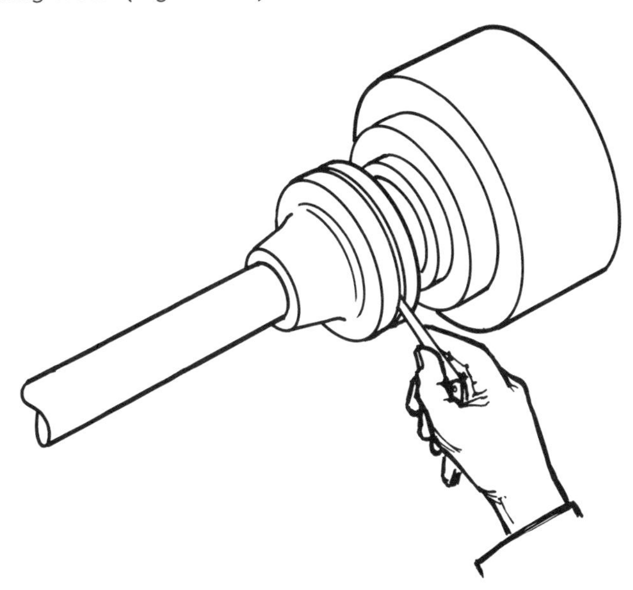

Figure 10.7 *Checking gearbox coupling. Support the propeller shaft close to the coupling to allow for bending, also deflection of the stern tube bearings.*

Exhaust System

If you have been plagued by water entering the exhaust pipe when you are reversing or running with a following sea, you can improve the system. A simple transom flap may be all that is necessary, and you can of course fit this while the boat is in service. However, if the engine level is really too low in relation to the waterline for a simple exhaust pipe run, conversion to a *waterlock* system, as shown in Figure 10.8, will get you out of trouble; but it must be installed correctly and be of adequate size for your engine. Install strictly in accordance with the supplier's instructions – some points to watch for are indicated in the diagram.

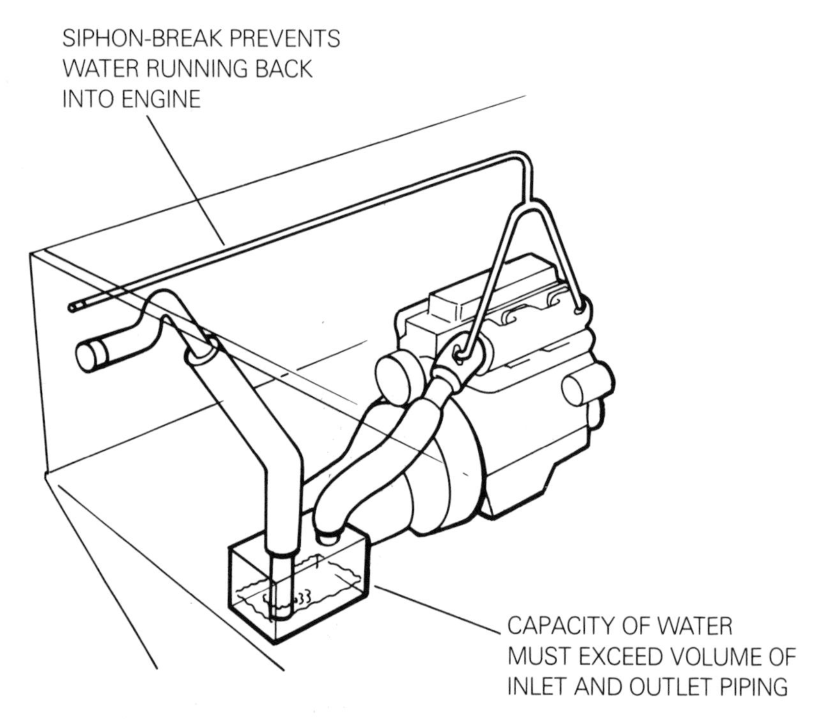

Figure 10.8 *Waterlock exhaust system*

Corrosion

Boat fittings and fastenings – as well as the hull itself when made of metal – may deteriorate through corrosion, and if you have suffered from this problem it will doubtless figure on your list of end-of-season jobs.

Corrosion prevention is recognized as one of the essential aspects of boat maintenance, and care is taken by boat-builders to avoid unsuitable metals and conditions which would cause corrosion. Things don't always work out as we would wish, however, so we shall examine reasons why corrosion occurs and what we can do about it.

Most corrosion on sea-going craft comes under the heading of *galvanic* – also called *electrolytic*. Electrolytic corrosion occurs when two 'dissimilar' metals are used for boat fittings or engine components that are immersed in sea water. They have also to be connected together electrically, which could mean being fitted to a metal structure such as the engine block or a steel hull; alternatively they may be 'bonded' by a pipe or other fitting inside the boat which is attached to both items. The conditions for electrolytic corrosion are shown diagrammatically in Figure 10.9. In effect we have a battery where the corroding metal is the *anode* as it loses ions in the electrolyte – i.e. the sea water. An electric current passes from the other metal, described as the *cathode*, via the hull or engine structure.

A metal corrodes more or less rapidly according to the potential or voltage difference between it and the 'dissimilar' metal with which it is associated. However, the relationship of the exposed surface areas of metal is also significant, and a large area of one material which is cathodic to a small anodic area will result in corrosion even when the metals have only a slight voltage difference – perhaps 0.1 volt.

If you have electrolytic corrosion you can deal with the problem by changing fittings to compatible metals – reasonably simple for pipe connections, bolts and other small components. However if you have a problem with propellers, stern-drive leg, shafts, rudders and other large expensive items it would be easier to fit sacrificial anodes, which will corrode away before the boat fittings. The anodes have to be correctly located and bonded; for best results it is

Figure 10.9 *Electrolytic corrosion*

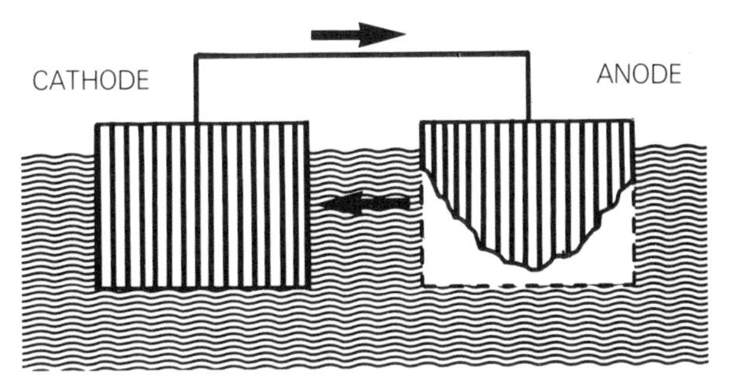

advisable to call in an expert to do the job. M. G. Duff & Partners are a very experienced company in this field. If the problem occurs on the engine it may be caused by a stray electric current – possibly a leakage to earth due to faulty insulation. The experts with their meters will track this down.

One problem resulting from the use of unsuitable metal for marine conditions is the *de-zincification* of brass, where sea water removes zinc from the alloy, leaving a weak shell of copper so that fastenings or fittings made from brass will snap off. Bronze is a much superior material.

Rusting of badly protected steel components is a well-known condition. In marine engines it is good practice to use stainless steel for components such as water-jacket plugs, exhaust pipe fittings and other thin-wall items. When this is not possible zinc or cadmium plating is often used; the latter is now considered injurious to health by some authorities and is less commonly used.

Pipes which are too small for the water flow will erode – particularly at the bends – until holes appear.

Instruments and Controls

Especially if your boat is several years old you can replace basic instrumentation by more sophisticated electrical systems which will monitor almost anything on the engine. A few years ago we were satisfied with an oil-pressure gauge and possibly a tachometer and water temperature gauge operated mechanically. Modern instrumentation is electrically operated, with sender units or switches screwed into the engine and transmission, wired up to the instruments in the control panel.

A second panel can be installed to monitor engine conditions on the flying bridge, for example. A warning panel which sounds a buzzer and flashes coloured lights to warn you of low oil pressure or high water temperature is available for most engines and can be retro-fitted to engines originally using only the basic instruments. Some of these devices are shown in Figures 1.41 and 1.42.

Stop-controls on boats with diesels vary from a length of cord passing from the fuel-pump stop lever through a hole in the engine box, to an electrically operated solenoid wired to a push-button or key-switch on the instrument panel. Most boat engines in fact have a push-pull stop cable, and if you want to add sophistication it is possible to fit a solenoid arrangement to most diesel engines. (Figure 10.10)

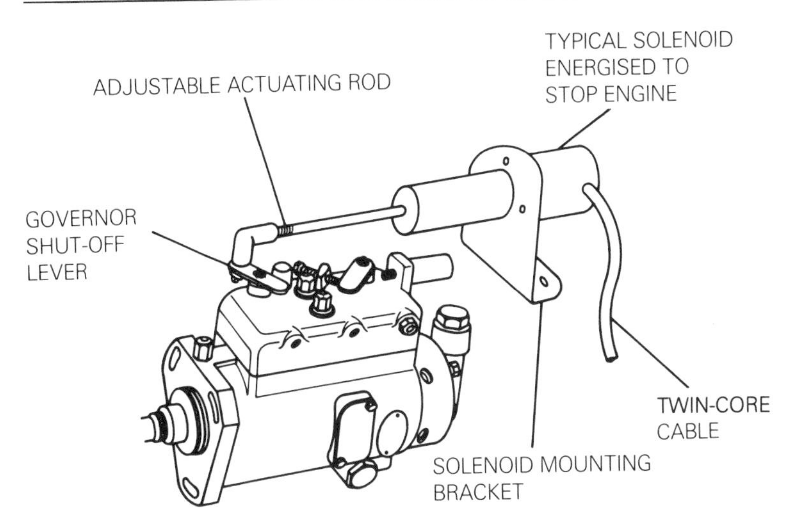

Figure 10.10 *Solenoid-actuated stop control*

Re-commissioning

This chapter has dealt with a few jobs to keep any boat-owner busy during the winter months, until it is time to relaunch and once more follow the preventive maintenance schedule through what I hope will be a trouble-free boating season.

Getting your boat ready for the new season will entail checking many items besides the engine installation. But leave yourself plenty of time to have a good look over the machinery – in particular the plumbing, which may have deteriorated since your last inspection, unless of course you dealt with this during the lay-up period.

Clean out the sea-water intake strainer, which may be located with the sea-cock or in a separate assembly between the sea-cock and water-pump inlet. Check the hose connections in the pipe and replace any that have cracks or have gone soft – you should be carrying spares in your engine's first-aid kit! Look for any doubtful rubber connections on the engine.

Check the exhaust system; any badly corroded piping as well as reinforced rubber tubing should be replaced. Hose clamps showing signs of corrosion should be replaced by stainless steel ones. Examine moulded rubber or metal silencers. Remove blanking plugs or tape and refit pipe to engine.

Turn off water drain taps and make sure all plugs are tightened; refit the water-pump impeller, or replace it by a new one if the blades look worn or have a permanent set. Use a new joint for the cover plate, and don't forget to grease the inside of the pump body or coat it with glycerine.

Fill the header tank with fresh water. Remove blanking tape from breather pipes, fuel-tank, air-cleaner etc. Replace the carburettor, not forgetting the flange joint, and connect up the controls and fuel supply pipe.

Drain inhibiting oil from the fuel-tank (diesel) and fill with fresh fuel; fit a new fuel filter element and bleed the system. Check the oil level in sump and transmission. Replace battery and leads; check the state of the charge with a voltmeter or hydrometer. Put on charge if necessary. Wipe excess petroleum jelly from terminals, starter and alternator.

Fill the stern-tube greaser or oil reservoir and lubricate the gland. If lubrication is by water feed, check the supply pipe from engine or scoop. Refit the propeller-shaft coupling, checking its alignment with the gearbox flange.

Check the alternator/water-pump driving-belt tension. It should not have deteriorated since your last routine check, but replace if there are signs of wear or cracks in the surface.

Start the engine and run it to make sure all systems are functioning correctly as indicated by the instruments.

Index

Page numbers for illustrations are in italics